HARMFUL FRESHWATER LIFE

Author ———

Ramasamy Santhanam, Ph.D. is the Former Dean of the Fisheries College and Research Institute, (presently Tamil Nadu Dr. J .Jayalalithaa Fisheries University), Tamil Nadu Veterinary and Animal Sciences University, India. He has served as a fisheries expert for various Government and nongovernment organizations in India and abroad. Dr. Santhanam has so far published 45 books on various aspects of Fisheries Sciences and Marine Biology; and his 2 books have been ranked as "Best Marine Biology Books of All Time" by the World Book Authority. On invitation, Dr. Santhanam participated and presented research papers and chaired sessions in 13 International Conferences held at Canada, Panama, The Netherlands, Switzerland, Spain, Germany, Denmark, Japan, South Korea, and Australia. He was a member of the American Fisheries Society, United States; World Aquaculture Society, United States; Global Fisheries Ecosystem Management Network, United States; and the IUCN's Commission on Ecosystem Management, Switzerland.

HARMFUL FRESHWATER LIFE

– *Author* –

Ramasamy Santhanam

ASTRAL

Publishing Authors Across the Globe

2025

Daya Publishing House®

A Division of

Astral International Pvt. Ltd.

New Delhi – 110 002

ISBN: 978-93-5919-548-3

Published by : **Daya Publishing House®**
A Division of
Astral International Pvt. Ltd.
– ISO 9001:2015 Certified Company –
4736/23, Ansari Road, Darya Ganj
New Delhi-110 002
Ph. 011-43549197
e-mail: info@astralint.com
Website: www.astralint.com

Printed at : **Replika Press Pvt Ltd**

PREFACE

Harmful freshwater algal blooms especially blue-green algal blooms due to cyanobacteria are the most notorious and problematic freshwater group. They are known to cause human and animal poisonings through liver and neurotoxicity in several countries such as America, Europe, Australia, Japan, Israel and Hungary. Freshwaters impacted by these blooms are off-limits to recreational uses such as bathing and fishing. Further, some bloom-forming species produce toxic compounds which are toxic to the other associated freshwater biota such as zooplankton, finfish and shellfish and vertebrate consumers of drinking water including humans. It is known that at least 49 human fatalities in a hemodialysis clinic from Brazil due to the consumption of microcystin- contaminated municipal water. Similarly 148 humans were hospitalized with hepatoenteritis (due to the cyanotoxin, cylindrospermopsin) in Queensland, Australia. Apart from these algal blooms, several groups of freshwater life such as toxic plants, protozoans, sponges, trematodes, annelids, spider, molluscs, fishes, amphibians (toads and frogs) , reptiles (water snakes, alligators and crocodiles) and mammals have also been found to be harmful either directly or indirectly due to their venomous stings or toxic tissues. All such species of animals and plants are not only affecting the food chain but also the fisheries of lotic and lentic freshwater systems leading to the adverse conditions of the economics of the countries. It is now felt that a thorough knowledge of harmful freshwater life is the time of need to safeguard our freshwater ecosystems.

Though several books are presently available on the biology and ecology of freshwater life, a comprehensive volume dealing with their harmful aspects has not so far been published. This publication titled "Harmful Freshwater Life" the first of its kind , would answer this long felt need. It deals with the biology and ecology of harmful freshwater life and their impacts on the health of the humans. It is hoped that the present publication when brought out would be of great use as a standard text-cum-reference for teachers, students and researchers of various disciplines such as Limnology, Aquatic Biology, Life Sciences, and Environmental Sciences; as a valuable guide to physicians, environmentalists, tourists, divers and fishermen and others involved in recreational activities; and as a valuable reference for libraries of colleges and universities.

I sincerely thank all my international friends who were very kind enough to share their images for the present publication. The services viz. photography and secretarial assistance rendered by Mrs. Albin Panimalar Ramesh are gratefully acknowledged.

Ramasamy Santhanam

CONTENTS

CHAPTER 1

INTRODUCTION

In the world, there are 5 major types of biomes viz. forest biome, desert biome, tundra biome, grassland biome and aquatic biome. Each biome is made up of a wide range of habitats. The aquatic biome is again subdivided into marine biome and freshwater biome based on the salt content of the water and the type of plants and animals present . Freshwater biomes are very important to human life because they provide us various services which may be of 4 types viz . supporting, provisioning, regulation and cultural. Supporting services include nutrient cycling, soil formation, and primary production; provisioning services include food, freshwater, electricity (hydroelectric power), wood and fiber, and fuel; regulatory services include climate regulation, flood regulation, disease regulation and water purification; and cultural services include aesthetic, spiritual, educational and recreational (Anon., 2009) . The wetlands which are also an important type of freshwater ecosystem serve as ideal habitats for several species of plants and animals; help control floods; and provide food for humans. Further, as these ecosystems are well connected to each other, they provide essential migration routes for many species of animals in their reproduction. Keeping these in view, these ecosystems have also been designated as "biological assets"

An Overview of Freshwater Ecosystems

Though most of the earth surface (about 70%) is covered with water, freshwater constitutes only 3.5 % and the rest (96.5%) is saltwater which is in the form of seas and oceans. The freshwater resources may vary in size from very small to large and are in the form of lakes, rivers, streams, ponds and ice. These freshwater bodies are home for about 40 % of the world's fish species and form a primary source of water supply for human beings due to their poor salt content which is less than 1% .Compared to the marine ecosystems, the freshwater ecosystem which supports various species of plants and animals is less extensive as far as the biodiversity is concerned (Anon., 2020).

A) Ecological parameters

1. **Climate :** Due to the uneven distribution of water in the various biomes of the world, the climate of the latter differs depending on their locations. In the summer, the average winter temperatures in freshwater biomes range from 65 to 75 degrees Fahrenheit. On the other hand, during winter , it is between 35 and 45 degrees Fahrenheit. Further, the climate of these biomes may also depend on water depth, season and location. Furthermore, unlike streams or rivers, in the freshwater biomes such as lakes and reservoirs, temperatures may also drop with their deepening of the water column and associated poor water movement.
2. **Precipitation:** The rainfall or precipitation in the freshwater biomes largely depends on the region where they are located . For example, as most wetlands are located in the temperate zones of the world they receive a lot of rainfall . On the other hand , the lakes and rivers which are located in mountains and valleys receive less rainfall. On an average, precipitation in the freshwater biomes may range from 25 to 200 cm per year.

B) Types of Freshwater Ecosystems

The freshwater resources differ in their movement. While some freshwater bodies like rivers and streams move continuously (lotic freshwater systems) other resources such as reservoirs and ponds are stagnant (lentic freshwater ecosystems)

1. **Lotic Freshwater Ecosystem:** Though the lotic freshwater ecosystems like river and streams move in one direction, some of these water bodies may flow from their origin and finally end at the mouth of the seas and such water bodies may be termed as estuaries, the salt content of which may vary from 1 to 3.5% (euryhaline) . Each lotic freshwater ecosystem is divided into 3 zones viz. source zone, transition zone and floodplain zone.
   a. **Source zone:** It is the narrow and fast-moving zone of a lotic freshwater ecosystem. Other characteristic features of this zone are : very cold temperature; high dissolved oxygen content; and poor primary, secondary and tertiary production. Freshwater fishes like trout are common in this zone.
   b. **Transition zone:** It is the middle part of the rivers and streams or river. This zone has gentle water flow and is deeper compared to the surface zone. Other characteristic features of this zone are its fairly high temperature , less oxygen content, higher productivity associated with more nutrients and high species diversity of plants and algae
   c. **Floodplain zone:** It is also called as the mouth of a river or stream. This zone is characterized by its slower flow, warm temperature, and very low dissolved oxygen content. This zone is also known for its murky water which is due to the movement and mixing of sediments in this zone

2. **Lentic Freshwater Ecosystem:** The size of lentic freshwater ecosystems may vary from a few square meters (ponds) to thousands of square km (lakes and reservoirs). Compared to the lotic ecosystems. These lentic systems are known for their poor productivity and species diversity

   a. **Littoral Zone:** This is the upper most zone near the shoreline of a lake or pond and is also the warm shallow zone of a lentic biotope. This zone which harbors a number of species of algae, but few species of plants, crustaceans, insects, clams, snails, and amphibians serves as the food source for ducks and turtles.

   b. **Photic zone or Limnetic zone** : This open water zone where adequate sunlight is available for pjhotosynthetic processes is the zone of a lentic ecosystem; and is rich in both phytoplankton and zooplankton. This zone serves an important role in the food chain of any freshwater ecosystem.

   c. **Aphotic zone of Profundal zone:** This is the deepest and cold part of the freshwater ecosystem where sunlight penetration is minimal. As photosynthesis is not possible in this zone , the animals of this zone are heterotrophs as they feed mainly on dead organisms.

3. **Wetland Freshwater Ecosystem:** The wetland is a stagnant water body which harbors mainly vascular plants. The swamps, marshes and bogs are stagnant water bodies and are called wetland ecosystems which are much productive due to their proximity of soil and water . The vascular plant species of these wetlands are called as hydrophytes as they are highly adapted to the humid and moist nature of these regions. Most common hydrophytes of these wetlands are pond lilies, cattails, sedges, tamarack, black spruce, etc.

C) Fauna and Flora of Freshwater Ecosystems

The freshwater biomes which make up 0.8% of the entire water body of the planet earth serves as ideal habitats for various species of fauna and flora. While some animals of freshwater dwell in lotic water bodies such as streams and rivers, some others inhabit stagnant water bodies such as lakes, ponds, and wetlands.

1. **Plants of freshwater biomes:** The plants of freshwater biomes are often influenced by factors such as water quality, sunlight, temperature , depth of water, rate of water flow etc.

   a. **Plants of streams and rivers:** In these freshwater biomes, the plants such as star grass, contrails, and tape grass are commonly seen on the edges of the water bodies. On the other hand, trees such as willows, cottonwoods, and river birch are seen in the shallow areas influenced by feeble water currents.

   b. **Plants of wetlands :** In the wetland areas such as bogs, swamps, marshes and ditches , the water is almost saturated throughout the year. Owing to this, these areas always muddy . Common plants of these wetlands include cattails and duckweeds; and tree species such as Cyprus , tamarack and black spruce are more prevalent in these biotopes.

c. **Plants of lakes and ponds:** These biomes do not grow a large community of plants owing to their greater depths. Plants which are thriving in these biomes include grasses and weeds. However, plants like cattails and lily pads may also be seen floating in the shallow lakes and ponds. Apart from these, the microalgae phytoplankton may also thrive in these biomes

2. **Animals of freshwater biomes:** The various freshwater biomes are known to support a significant diversity of animals which possess special adaptations and defense mechanism to survive and flourish in these biotopes. The most common animals of these biomes are given below.

   a. **European eels:** The European eels are mostly seen in seas and rivers. They possess a sleek-like skin to glide easily through the water column. Their chief food items include invertebrates, small fish, and insect larvae.

   b. **Black piranha:** It is mostly seen in rivers. It possesses sharp teeth and is very ferocious. It feeds largely on fish, beef hearts, decaying carcasses, seeds and fruits.

   c. **Frogs:** These incredible jumpers are uniquely adapted to the freshwater biomes particularly in trees, creeks and ponds. Their food consists of grasshoppers, moths, mice, mosquitoes, and dragonflies

   d. **Crocodiles:** The crocodiles of freshwater biomes are known for their characteristic sharp teeth and strong claws. They are commonly seen in canals, creeks, ponds and wetlands. They feed mostly on fish, mammals, invertebrates , other reptiles, and birds.

   e. **Freshwater turtles:** The major habitats of these animals include rivers, lakes, and ponds. Their strong shell protects them from predators. These turtles consume frogs, fish, insects, and certain plants.

   f. **River otters:** These otters possess long tails which serve as oars to easily glide through the water column. Their major habitats include rivers, creeks and streams. They consume fish. frogs, turtles and water birds. It is worthy of mention here that the North American otters are semi-aquatic living both in water and land.

   g. **Water birds:** Some birds like ducks and geese are also thriving in these freshwater ecosystems viz. lakes and rivers. However, some birds like swallows thrive only in ponds and swamps. For the insect-eating swallows, the swamps form ideal shelters .

   h. **Insects:** Freshwater biomes possess various species of insects such as bees, wasps , mosquitoes, dragonflies, water spiders, etc.

The other animals of freshwater biomes include Amazon River dolphins , Clear Lake Hitch, raccoons, etc.

D) Harmful freshwater life

The development and proliferation of hazardous plants and animals in the different aquatic systems may also largely be due to the prevailing adverse water conditions. Though several groups of freshwater life have been found to be inimical to human life and associated boating, diving and recreational activities, harmful algal blooms and certain categories of fishes (like catfishes, pufferfishes, stingrays and electric fishes) and toads and frogs have been considered to be a menace.

1. **Harmful algal blooms:** Freshwater algal bloom formations create health hazards for humans through the production of toxins. Majority of these algal bloom problems reported worldwide are largely due to the microalgae particularly the blue- green algae or cyanobacteria. Harmful cyanobacteria produce neurotoxic, hepatotoxic and dermatotoxic compounds. Acute exposures of these compounds have been reported to be harmful to humans resulting in dermatitis, respiratory illnesses, cancer, and even death. In addition to human health problems, these algal blooms may pose serious threat to fisheries and aquatic environment. It is worth mentioning here that a cyanobacterial density of > 100,000 cells per ml may make its habitat unfit for human activities including swimming. It appears that liver-toxic (hepatotoxic) microcystin-containing cyanobacterial freshwater blooms have been reported from all continents and almost every part of the world. While mass occurrences of neurotoxic cyanobacteria are common in North America, Europe and Australia, where they have caused several animal poisonings, blooms which have caused both liver and kidney damage due to the toxin cylindrospermopsin have been reported in Australia, Japan, Israel and Hungary (Sivonen and Jones,1999).
2. **Harmful freshwater animals:** Freshwater pufferfish are believed to be the second most poisonous vertebrate in the world . Almost all pufferfish have been reported to contain tetrodotoxin, a substance 1,200 times more poisonous than cyanide. About 50 deaths from pufferfish poisoning are reported in Japan annually. Among the venomous freshwater fishes, the stingrays are very notorious ones. Potamotrygon rays which have robust tails with stingers in mid-distal position have been reported to cause injuries throughout the year in several freshwater systems including Amazon and Paraguay Rivers. Among the higher vertebrates, the anuran amphibians, especially the typicalatoadsa(genus *Bufo*) are known for their hazardous nature in freshwater bodies. Their skin secretions contain a moderately potent poison viz. bufotoxin which has several components including bufagin, bufotenine and serotonin. Bufagin has effects on the heart similar to those of digitalis; abufotenine, a hallucinogen; and have aserotonin, a vasoconstrictor.

E) Monitoring of freshwater ecosystems and their life

Unfortunately, urbanization, industrialization and other anthropogenic activities exert drastic pressure on these freshwater ecosystems. Pollutants and associated water contaminations in aquatic ecosystems have long-term impacts on their associated biota. By recognizing the need for reducing the pollution loads, we can maintain or restore freshwater ecosystems to a sustainable state which will help native aquatic species to flourish. In order to safeguard these valuable ecosystems viz. rivers, streams, lakes and ponds, it is also very important to form stream team, or community group with volunteers These people could work together to monitor and restore the freshwater biomes from further destruction. It is also worth mentioning here that, a number of plant and animal species have been reported to be harmful causing hazards in the day to day life of humans. In order to utilize the services of healthy freshwater ecosystems and their biodiversity , it is very much needed to have a thorough knowledge on the characteristics, types and fauna and flora of freshwater ecosystems in general and the harmful freshwater life in particular.

CHAPTER 2
HARMFUL FRESHWATER ALGAE

2.1. Toxic Freshwater Microalgae

2.1.1. Blue-green algae (Cyanobacteria: Cyanophyceae)

2.1.1.1. Cyanobacterial Toxins

The toxins produced by the cyanobacteria are hepatotoxins, neurotoxins, cytotoxins and dermatotoxinsa

i) Hepatotoxins

Hepatotoxins are of two kinds viz. microcystins and nodularins which are cyclic peptides. In cases of intoxications of hepatotoxins, internal bleeding and liver damage may occur. Characteristic symptoms of intoxications of these toxins are gastrointestinal and liver disorders, weakness and anorexia. Hepatotoxins have also been reported to induce apoptosis and necrosis of hepatocytes and are also promoters of cancerous processes (Jakubowska aandaSzel g-Wasielewska, 2015). While the microcystins are seen in the freshwater cyanobacteria, the nodularins are found in the species of *Nodularia* which are known to occur only in the brackish or salinic waters (https://en.wikipedia.org/wiki/Nodularia).

Microcystins : The microcystins which are hepta peptides (seven amino acids) have been isolated in >80 structural variants. Toxicity studies made with mice have shown that the LD_{50} (Lethal Dose) of this toxin was 50 µg/kg body weight (https://www.sciencedirect.com/topics/nursing-and-health-professions/nodularin).

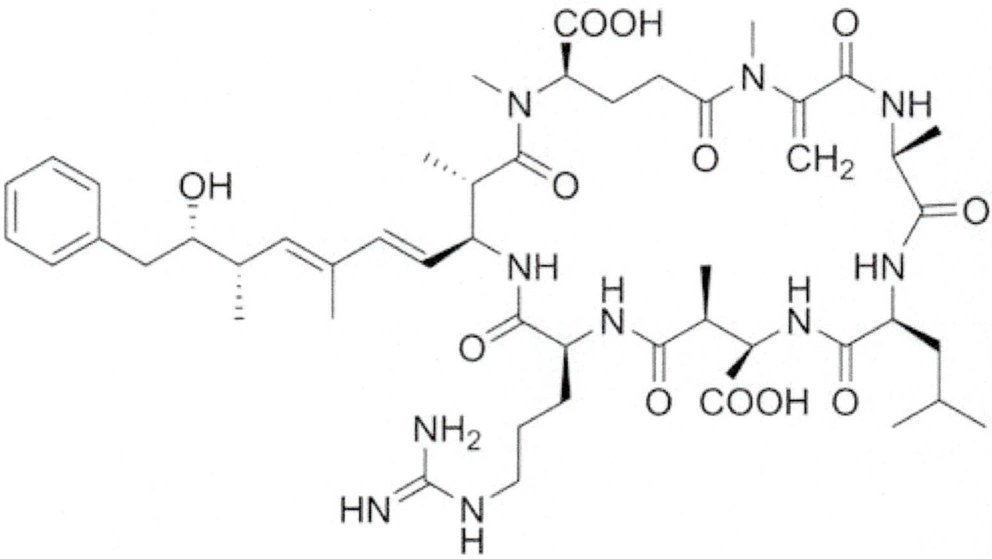

Microcystin

Microcystin producing Cyanobacteria : Cyanobacteria, as part of a larger algal community, form the base of the food web in many aquatic environments (both marine and freshwater). The cyanobacterial bloom species which produce microcystins are of picoplanktonic cell size (0.2 to 2.0 µm). These microcystins are most commonly produced by the genus *Microcystis*, although other genera such as *Planktothrix, Oscillatoria, Anabaena, Anabaenopsis, Nostoc, Hapalosiphon, Snowella,* and *Woronichinia* are also known to produce this toxin (Marmen et al., 2016; JakubowskaaandaSzel g-Wasielewska, 2015)

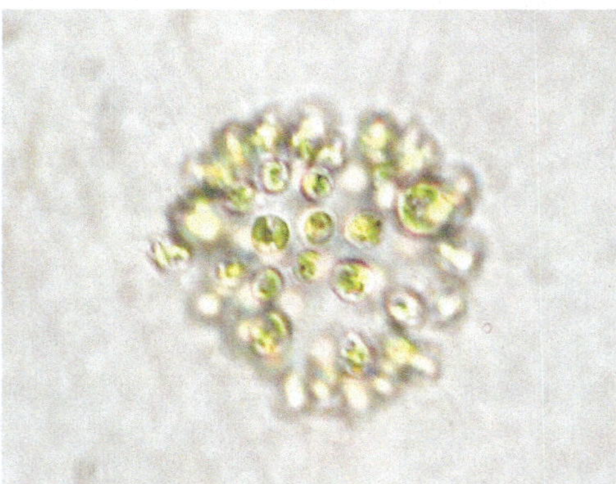 

Microcystis aeruginosa *Anabaena flos-aquae*

Image Credit:.Kristian Peters , Wikimedia; Environmental Protection Agency, Wikipedia

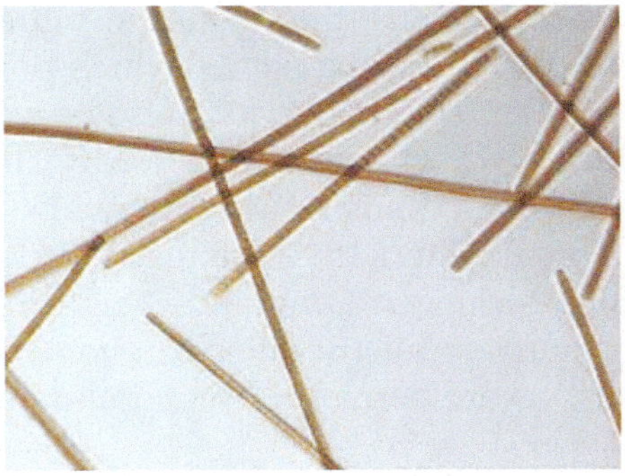

Planktothrix agardhi
Image credit: Nordic Microalgae

Nostoc sp.
Yamamaya, Wikipedia

Global distribution of microcystin producing species: The microcystin producing species are found distributed globally from polar regions to the tropics. The toxic *Microcystis* has been reported from Australia, Brazil, China, Portugal, Sweden and the USA. In the UK, apart from *Microcystis* spp., other cyanobacteria genera such as *Oscillatoria, Plantothrix, Anabaeana, Pseudoanabaena, Aphanizomenon, Snowella* and *Gomphospaeria* have also been recorded. Among the harmful bloom forming cyanobacterial species, *Microcystis* aeruginosa has been reported to be most common toxic species in eutrophic freshwater and is of economic and ecological importance in many countries (Turner et al., 2018).

Habitat of microcystin producing species: The blooms of microcysin producing species are commonly found in in eutrophic freshwater systems. Local environmental factors, including phosphorus and ammonia concentrations and pH are normally responsible in the development of these toxic blooms. Marmen et al.(2016) reported that increases in nutrient load, temperature, salinity, and UV light may all contribute to the emergence of microcystin-producing harmful algal blooms .

Hazards of microcystins: The concentration of toxic microcystins in the bloom forming species of cyanobacteria has been reported as 25 mg per dm^3 of water and 20 mg per gram of dry matter (Jakubowska aanda Szel g-Wasielewska, 2015). Cell densities above 20,000 cells/mL may normally contain microcystins above the 20 μg/L . The guideline value of microcystin-LR for drinking water however, is 1μg/l (WHO,1998). These harmful cyanobacterial blooms are one of the major threats to water quality worldwide, affecting many ecologically and economically important water bodies, and occasionally causing severe health problems and/or mortality in livestock and humans. In the United States alone, these blooms have been estimated to cause a loss of $2 billion per year by making the water bodies unfit for recreation, drinking or agriculture.

Health problems due to microcystins: In humans, the organ most affected by the microcystins is the liver, and specifically the hepatocytes which possess the target protein phosphatases type 1 and 2. Damage to liver may also be caused by microcystins as a result of an extended level of γ-glutamyl transferase(Chorus and Bartram, 1999). High concentrations of microcystins in water may also lead intoxication and death in animals and humans, and they are also said to be potential carcinogen (Marmen et al., 2016). Esterhuizen-Lond and Pflugmacher (2020) reported that acute and chronic exposures to microcystins can be harmful to humans resulting in respiratory illnesses, severe dermatitis, mucosal damage, cancer, organ failure and death

- **Short Term Health Effects:** Gastrointestinal, liver inflammation, and hemorrhage and liver failure leading to death, pneumonia, dermatitis (Lopez et al., 2008)
- **Long Term Health Effects**: Tumor promoter, liver failure leading to death (Lopez et al., 2008)

ii) Neurotoxins

These cyanotoxins are alkaloids which include anatoxins viz. anatoxin-a, with derivatives, anatoxin-a (s), saxitoxin with analogues and β-N-methylamino-L-alanine (BMAA)

a) Anatoxins: The anatoxins include the bicyclic, secondary amine alkaloid viz. anatoxin-a and its analog, homoanatoxin- a. Both molecules share almost identical toxicological properties.

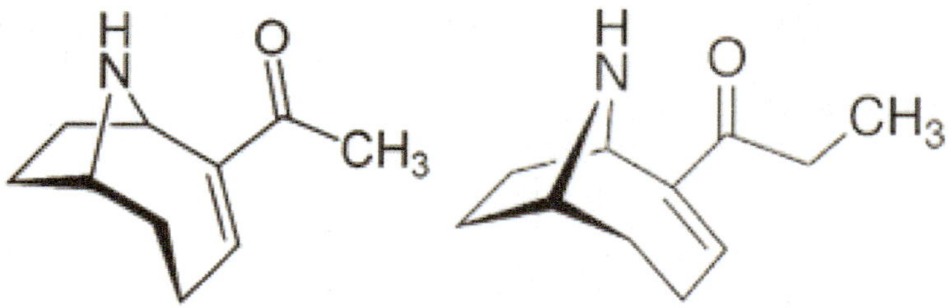

Anatoxin-a Homoanatoxin- a

Anatoxin (ATX) producing cyanobacteria:. ATX is produced by a variety of cyanobacteria species such as *Anabaena flos-aquae, Anabaena lemmermannii, Chrysosporum (Aphanizomenon) ovalisporum, Cuspidothrix issatschenkoi (=Aphanizomenon* issatschenkoi), *Cylindrospermopsis* sp., *Raphidiopsis mediterranea, Cylindrospermum* sp., *Dolichospermum (Anabaena)acircinale), Oscillatoria* sp., *Planktothrix* sp., *Phormidium* sp., *Tychonema* sp.,*Microcystis* sp. and *Woronichinia* sp..

***Phormidium* sp.**
Image credit: Gabenitez, Wikimedia Commons

Global distribution of anatoxin producing species: Temperate regions of Europe, the USA, Japan and New Zealanda

Habitat of anatoxin producing species: ATX occurs at high concentrations in waterbodies influenced by human activity, i.e. effluents from wastewater or run-off from agricultural areas. High concentrations have been observed in beached or floating benthic mat material or macrophytes to which ATX-producing cyanobacteria are associated. It is also accumulated in fish

Hazards of anatoxin : Its values of concentration detected in normal water samples, surface blooms and in fish were found to be > 4 µg/L to, > 1000 µg/L and <50µg/g respectively (WHO,2019). Animal fatalities have been reported due to its ingestion.

Effects of anatoxins on human: WHO (2019) reported on the suspected human poisoning by ATX occurred in a 17-year-old boy who died 2 days after swallowing water while swimming in a golf course pond containing an *Anabaena flos-aquae* bloom. Further 3 ATX associated disease outbreaks with neurological symptoms have also been recorded

Neurological effects of anatoxins: ATX has been reported to affect nerves and muscles. Two daily intraperitoneal injections of 0.21 mg ATX/kg in mice caused decreased motor activity, altered gait, difficulty breathing and convulsions 5–6 minutes postexposure and finally death within 10 minutes. ATX is also known to affect other nicotinic cholinergic receptors in the cardio-vascular system to increase blood pressure and heart rate, and in the brain (WHO, 2019)

Patocka, et al. (2011) reported that anatoxin was a potent acetylcholinesterase inhibitor at the nerve synapse, which results in paralysis, asphyxiation, decreased movement, exaggerated abdominal breathing, cyanosis, convulsion, and ultimately death.

Effects of Homoanatoxin-a : It displayed similar toxicological properties as anatoxin-a (WHO,2019)

b) Saxitoxins: Saxitoxins (STX) are a group of carbamate alkaloid neurotoxins and were originally isolated from marine shellfish where they are concentrated from red tide causing marine dinoflagellates . Consumption of affected shellfish have resulted deaths in humans. There are different types of saxitoxins such as anon-sulfated STX and neosaxitoxin (neoSTX); mono-sulfated gonyautoxins (GTXs 1–6), or di-sulfated (C1-4 toxins). In addition to these , decarbamoyl variants of these analogs such as decarbamoyl-saxitoxins (dcSTX, dcneoSTX), decarbamoyl-gonyautoxins (dcGTXs 1–4), and 13-deoxy-decarbamoyl derivatives (doSTX, doGTX 2,3) also exist (Wiese et al., 2010)

Saxitoxins of cyanobacterial species

Name of STX	Cyanobacterial species
Saxitoxin	*Aphanizomenon flos-aquae, Anabaena circinalis , Cylindrospermopsis raciborskii* and *Scytonema crispum*
Gonyautoxins 1-6	*Anabaena circinalis* and *Scytonema crispum*
C1,& C2 toxins	*Anabaena circinalis*
Neosaxitoxin	*Aphanizomenon flos-aquae , Cylindrospermopsis* raciborskii and *Scytonema crispum*
decarbamoyl-saxitoxins	*Anabaena circinalis ,Lyngbya wollei* and *Scytonema crispum*
decarbamoyl-gonyautoxins 2,3	*Anabaena circinalis , Lyngbya wollei* and *Scytonema crispum*
L. wollei toxin (LWTX1-6)	*Lyngbya wollei*

Source: Wiese et al., 2010; Smith et al., 2012

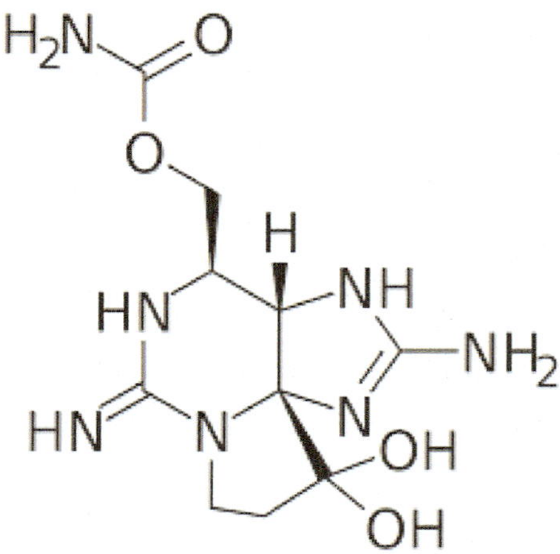

Saxitoxin

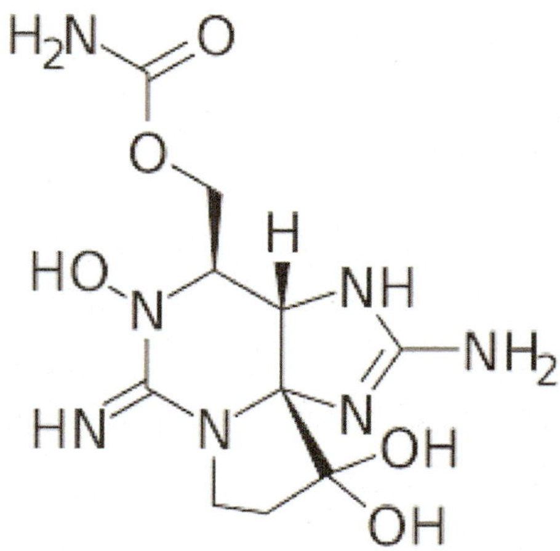

Neosaxitoxin

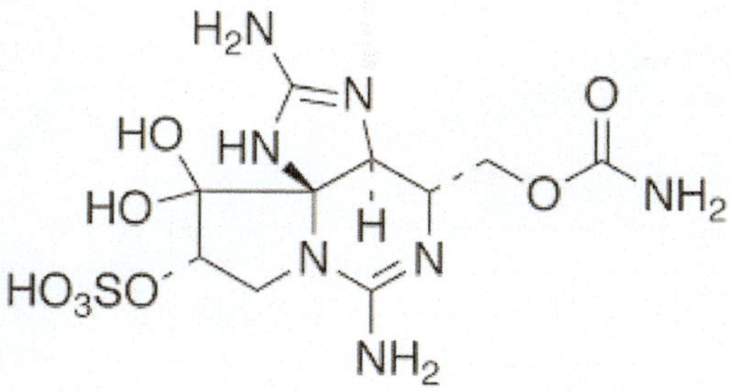

Gonyautoxin 1

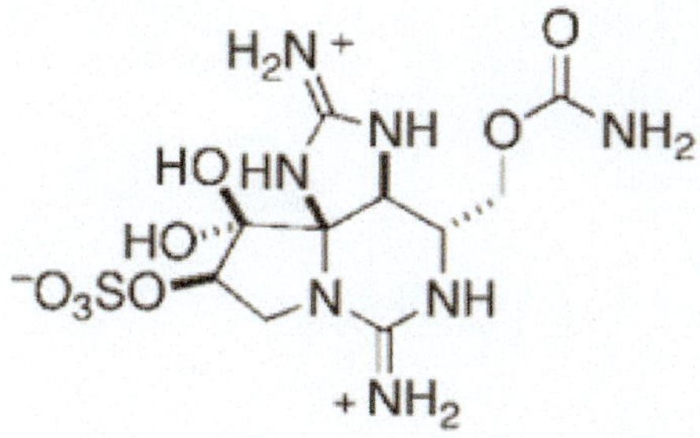

Gonyautoxin 2

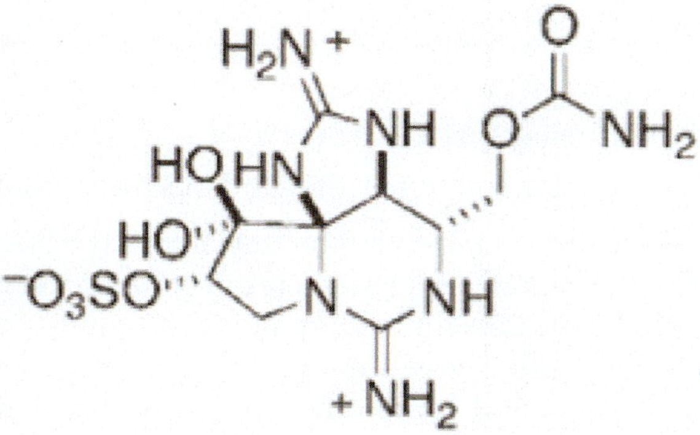

Gonyautoxin 3

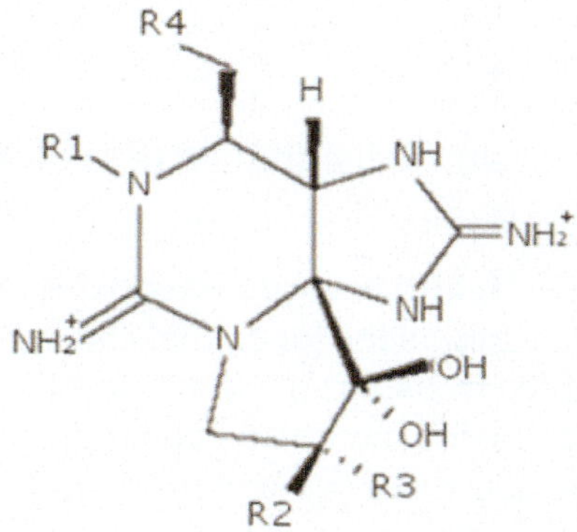

Gonyautoxin 4

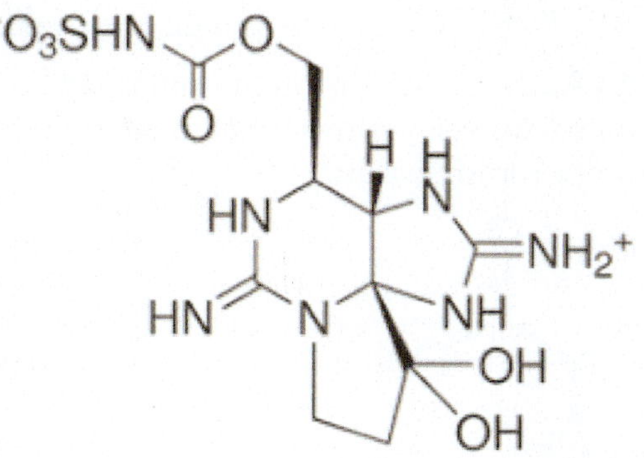

Gonyautoxin 5

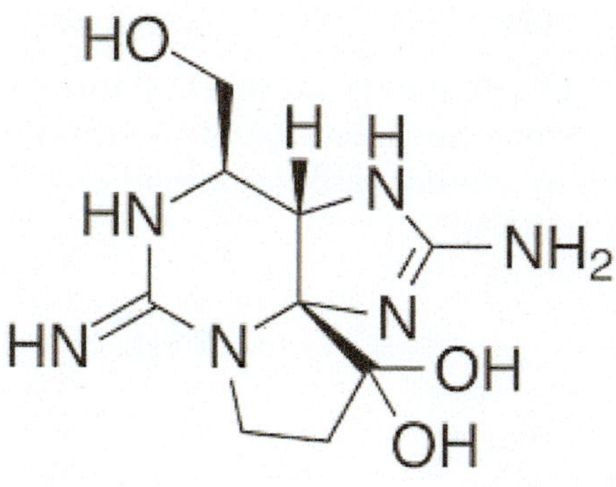

Decarbamoyl saxitoxin

Saxitoxin producing cyanobacteria: Saxitoxins have also been found in the cyanobacterial species such as *Aphanizomenon flos-aquae, Anabaena circinalis, Anabaena flos-aquae, Anabaena lemmermannii ,Lyngbya wollei , Scytonema acrispum, Planktothrix agardhii* and *Cylindrospermopsis raciborskii* (Chorus and Bartram, 1999)

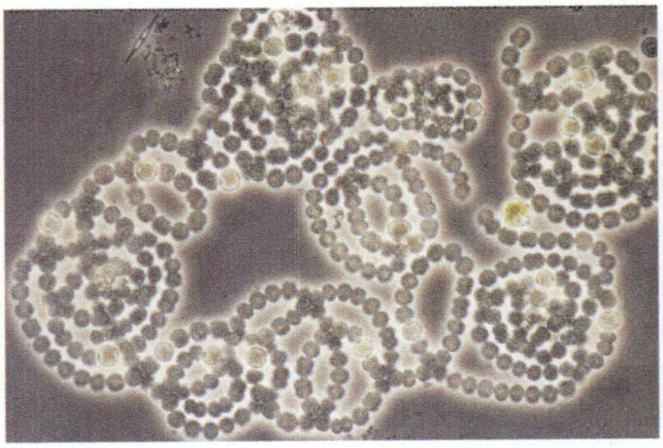

Anabaena circinalis

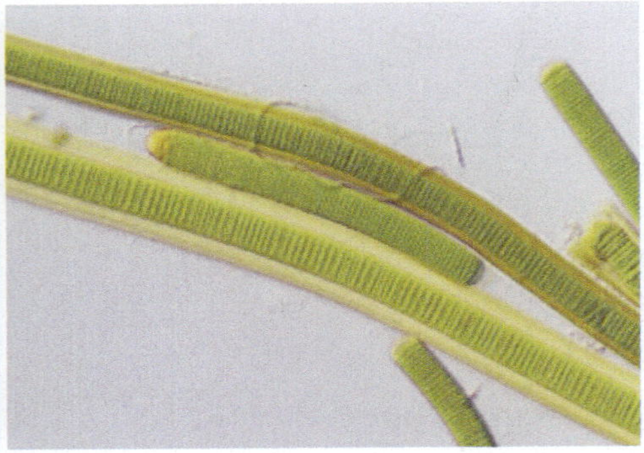

Lyngbya sp.

Credit for images: Wikipedia

Global distribution of saxitoxin producing species: Worldwide

Habitat of saxitoxin producing species: Both in freshwater and marine habitats

Health hazards: Saxitoxins are also called as paralytic shellfish toxins (PSTs) as they inhibit the functional ability of the nervous appendage as a result of the blockage of the transport of Na+anions through the sodium channels in nervous cells. These saxitoxins have also been reported to show dermotoxic and cytotoxic effects, such as skin rash, eye irritation, abdominal pains and symptoms similar to influenza. While the intraperitoneal dose LD_{50} for mice was found to be 5 µg/kg–1, the lethal dose for humans was 1 mg/kg–1a(Jakubowska aandaSzel g-Wasielewska, 2015)

Short term health effects in human : Tingling, burning, numbness, drowsiness, incoherent speech, and respiratory paralysis leading to death (Lopez et al., 2008)

c) β-N-methylamino-L-alanine (BMAA): It is a non-protein amino acid and is otherwise called as or L- -amino-β-methylaminopropionic acid, 2-amino-3-(methylamino)- propionic acid or β-diaminopropionic acid

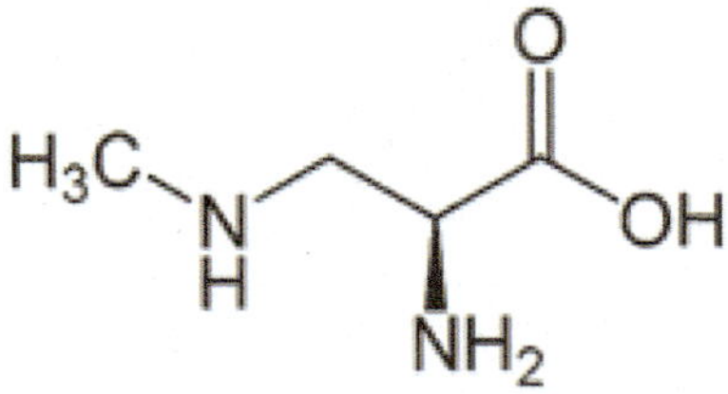

β-N-methylamino-L-alanine

BMAA producing cyanobacteria: *Microcystis* sp. , *Synechococcus* sp. , *Anabaena variabilisa, Nostoc* sp., *Planktothrix agardhiia* and *Cylindrospermopsis raciborskiia*

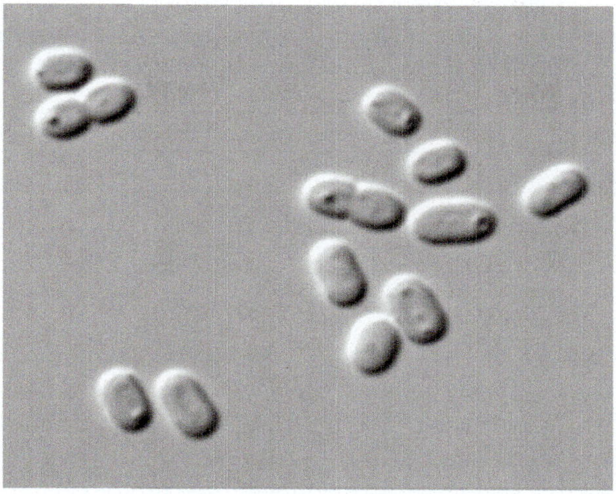

Synechococcus sp.
Image credit: Wikipedia

Global distribution of BMAA producing species: worldwide; USA, Australia, USA, Europe

Habitat of BMAA producing species: They are found as plankton in shallow water bodies

Concentrations of BMAA: The concentrations of BMAA in free living cyanobacterial organisms and other animals have been reported to range from below detection limit to about 7000 µg/g which is associated with a potential long -term health hazard for humans. *In vitro* studies, BMAA was found to be toxic to neurons at concentrations of 10-30 nM (Brand et al., 2010). The concentration of BMAA recorded in free-living cyanobacteria and animals are given below.

BMAA in free-living cyanobacteria

Cyanobacterial species/strain	BMAA (µg/g)
*Microcystisa*PCC 7806	4
*Microcystis*aPCC 7820	4
Synechococcus PCCa6301	25
*Planktothrix agardhii*aNIES 595	318
*Anabaena variabilis*aATCC 29413	35
*Cylindrospermopsis raciborskii*aCR3	6478
*Nostoc*aPCC 6310	42

Source: Cox,a et al., 2005

Concentrations of BMAA in animals in the Caloosahatchee River

Species	Tissue type	BMAA (ug/g)
Utterbackia imbecillis	Mantle muscle	251
Crassostra virginica	Mantle muscle	281
Amia calva	Tail	2559
Micropterus salmoides	Tail	2388
Atractosteus spatula	Head	2290

Source: Brand et al., 2010

Health hazards of BMAA: BMAA has been implicated as an important environmental risk in the development of neurodegenerative diseases such as Alzheimer's disease, Parkinson's disease, and Amyotrophic Lateral Sclerosis (ALS) which is otherwise called as Lou Gehrig's disease (Brand et al., 2010). Rastogi et al. (2015) reported that this toxin may be responsible for motor system disorder and it is a glutamate agonist. Further, it is known to , increase the intracellular concentration of calcium in neurons and induce neuronal activity by hyperexcitation.

iii) Cytotoxins

Cylindrospermopsin (CYN) : It is a guanidine alkaloid. Two structural variants of Cylindrospermopsin viz. 7-epicylindrospermopsin (7-epi-CYN), and 7-deoxycylindrospermopsin (7-deoxyCYN),have been identified. Among them, the former has a similar toxicity as that of the mother molecule, while latter's toxicity is still unknown. CYN is the second most studied freshwater toxin after microcystin and it poses more hazards to water users since the liberation of CYN to surrounding water is much higher compared to other toxins (Sotton et al., 2014).

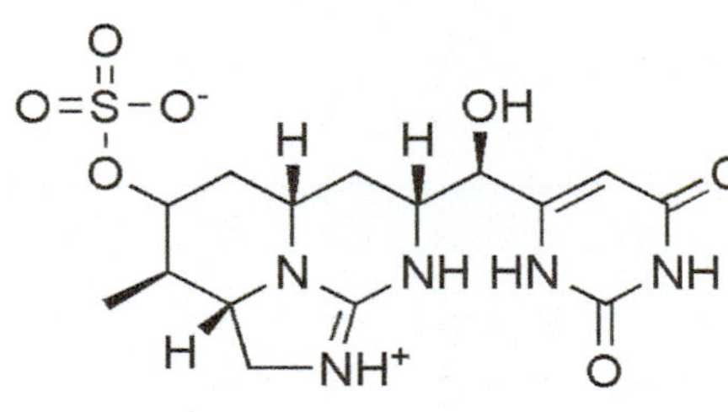

Cylindrospermopsin

Cylindrospermopsin producing cyanobacteria: *Raphidiopsis raciborskiia(= Cylindrospermopsis raciborskii), Chrysosporum ovalisporuma(=Aphanizomenon ovalisporum), Anabaena* sp., *Lyngbya wollei*, and *Umezakia natans*

Global distribution of Cylindrospermopsin (CYN) producing species: *Raphidiopsis raciborskiia(= Cylindrospermopsis raciborskii), Chrysosporum ovalisporuma(=Aphanizomenon ovalisporum* -Worldwide; *Lyngbya wollei*,- USA; *Umezakia natans*- Lake Mikata, Fukui, Japan

Habitat of cylindrospermopsin producing species: Lakes, reservoirs, streams, rivers, ponds and shallow systems.

Concentration CYN in freshwater ecosystems: It is detected at variable concentrations (from 0 to more than 800 lg l-1 total CYN) in samples from different freshwater ecosystems (Sotton et al., 2014)

Health hazards of Cylindrospermopsin: It is a "slow toxin", with an LD_{50} value of 2.1 mg kg–1a(after 24 h) and LD_{50} of 0.2 mg kg–1a(after 5–6 days). It has several paths of toxic activity viz. inhibition of glutathione, synthesis of proteins and cytochrome P450. These toxicities are associated with oxidative stress, leading to loss of chromosomes and splitting DNA, and disruptions of the cell cycle as a result of which apoptosis or necrosis of cells occur . Cylindrospermopsin has also been reported to cause liver and kidney damage as well as pathological changes of lungs, heart, stomach, adrenal glands and of the vascular and lymphatic systems (Jakubowska aandaSzel g-Wasielewska, 2015)

Short term health effects: Gastrointestinal, liver inflammation and hemorrhage, pneumonia and dermatitis (Lopez et al., 2008)

Long term health effects: Malaise, anorexia, and liver failure leading to death (Lopez et al., 2008)

iv) Dermotoxins (Dermatotoxins)

a) Lyngbyatoxins: Lyngbyatoxins are indole alkaloids. Three isoforms of lyngbyatoxin viz. lyngbyatoxins a, b and c have been identified. Lyngbyatoxin-a is identical to an isomer of teleocidin which is a tumor promoter isolated from *Streptomyces mediocidicus* ISP 5021

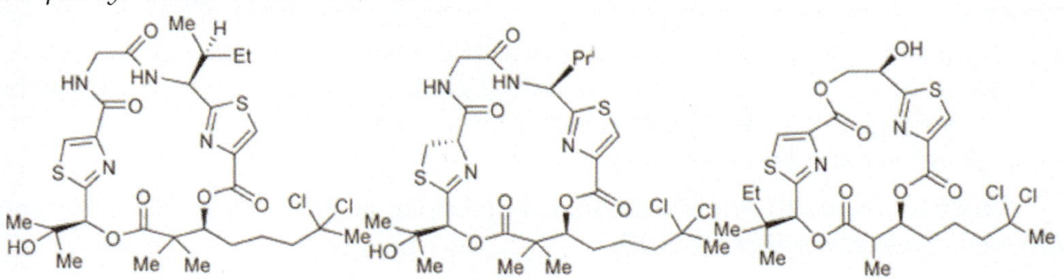

Lyngbyatoxins a-c

Lyngbyatoxin producing cyanobacteria: *Lyngbya wollei*

Global distribution of lyngbyatoxin roducing species: *Lyngbya wollei*ahas been found in southeastern USA. Genetically similar subgroups of this freshwater species have also been recorded from Laurentian Great Lakes and St Lawrence River of North America.

Habitat of lyngbyatoxin producing species: This filamentous species forms dark green to black mats on the bottom of rivers and lakes. Benthic mats however, remain inconspicuous until they float to the surface because of trapped gas bubbles. This taxon is found under contrasting environmental conditions viz. very clear, thermally and chemically stable, and heavily mineralized Florida Springs; and turbid, high dissolved organic C, and seasonally variable conditions as in the St Lawrence River.a

Impact on environment: Mats of *Lyngbya wollei* have been reported to provide shelter for invertebrates. St Lawrence River, and its wetlands which are dominated bya this species are invariably characterized by a lower biomass of invertebrates and large fishs. On the other hand ,replacement of macrophytes byа*L. wollei*аmats induced a shift in trophic structure and coincided with a decrease in carrying capacity for fish, and significantly altered the dynamics of freshwater ecosystems. (Hudon et al., 2014)

Health hazards of lyngbyatoxins

The toxicity of this toxin with mice (oral route) showed an LD_{50} value of 250 µg/kg (Rzymski and Poniedz,2012)

While lyngbyatoxin-a has a skin tumour promoting activity, lyngbyatoxin-b and lyngbyatoxin-c have shown 1/200th and 1/20th the activity of lyngbyatoxin-a, respectively . With regard to the health hazards of lyngbyatoxins on humans , at 6 % dose, these toxins showed slightly lipophilic, penetration through human skin after 1 h of exposure to 26 µg/cm^2 (Rzymski and Poniedz,2012).

The most common symptoms of lyngbyatoxins were associated with human skin and are usually defined as "seaweed dermatitis". The irritant properties of these toxins were also recorded with ears (sore, discharge) and eyes (sore). The most frequent general symptoms were however, headache, nausea, vomiting and diarrhea. Young adults showed the highest sensitivity to lyngbyatoxin. First symptoms occurred within a few minutes to a few hours after exposure. Visible dermatitis with redness occurred after 3-20 h and lasted for 2-12 days. Lyngbyatoxins may also be carried by wind and cause a health risk for humans if they are not even exposed to water (Rzymski and Poniedz,2012).

Short term health effects: dermatitis (Lopez et al., 2008) ; cytotoxic and gastroenteritis (Rastogi et al., 2015)

Long term health effects: Skin tumors (Chorus and Bartram, 1999)

b) Aplysiatoxin (AT) and c) debromoaplysiatoxin (DAT): These compounds belong to phenolic bislactones

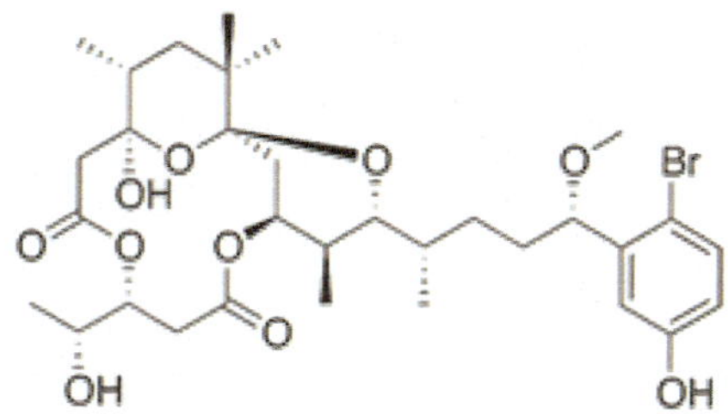

Aplysiatoxin

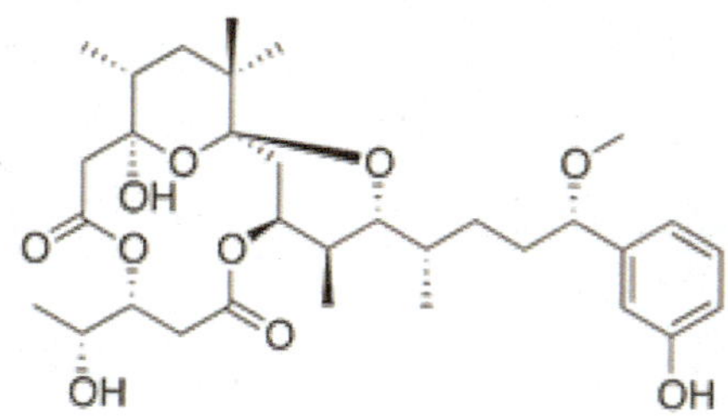

Debromoaplysiatoxin

AT and DAT producing cyanobacteria: AT and DAT are commonly seen in the species of *Lyngbya, Schizothrix* and *Planktothrix* . Both AT and DAT were first isolated from the marine mollusks of the genus Stylocheilus which feed upon these cyanobacterial genera

Global distribution of AT and DAT producing species: Worldwide

Habitat of AT and DAT producing species: All freshwater bodies

Health hazards of AT and DAT: Both AT and DAT have shown severe ear irritation in experimental mice. In human skin cells, these toxins have been reported to possess the inhibition of epidermal growth factor (EGF) (10 times higher for aplysiatoxin) and activation of ornithine decarboxylase. Further, these toxins displayed tumorigenic properties and were activators of protein kinase C. Furthermore, these toxins caused differentiation of HL-60 cells into macrophages (Rzymski and Poniedz,2012)

Topical application of debromoaplysiatoxin caused an irritant pustular folliculitis in humans and severe cutaneous inflammation in rabbits and in hairless mice (Rzymski and Poniedz,2012)

Direct contact of aplysia- and debromoaplysiatoxin contaminated water resulted acute skin irritation, rashes and blisters . Further , a risk of inhalation of aplysia- and debromoaplysiatoxin from aerosols transported by the wind has also been reported near the water bodies (Rzymski and Poniedz,2012).

d) Lipopolysaccharides (LPS): Lipopolysaccharides form complexes with proteins and phospholipids

Lipopolysaccharide producing cyanobacteria: *Microcystis aeruginosa*

Global distribution of lipopolysaccharide producing species: Worldwide

Habitat of lipopolysaccharide producing species: It lives near the surface of all freshwater ecosystems as it largely depends on oxygen

Health hazards of lipopolysaccharides: The toxicity of *M.aeruginosa* in experimented mice showed an LD_{50} value of 50 mg/kg. However, the hazards of this toxin on human health and its mechanism of action still remain unclear. This is largely due to the difficulties in distinguishing the symptoms caused only by LPS. However, the reported symptoms due to this toxin include skin and eye irritation, respiratory problems ,hay fever, , blistering of mucous membranes, headaches and dizziness and fever (Rzymski and Poniedz,2012).

Rastogi et al.(2015) reported that these endotoxins may harm immune and detoxification system; and cause irritation, and allergic effects. Further, according to them these toxins inflamed gastrointestinal epithelium and caused death due to its high enough concentrations in blood.

Short term health effects: Gastrointestinal; and dermatitis (Lopez et al., 2008).

v) Other cyanotoxins

a) Thionsulfolipid : The structure of this thionsulfolipid has been identified as 6-sulfo- -d-quinovopyranosyl- (1 1′)-2′-O-acyl-3′-O-thioacyl-d-glycerol

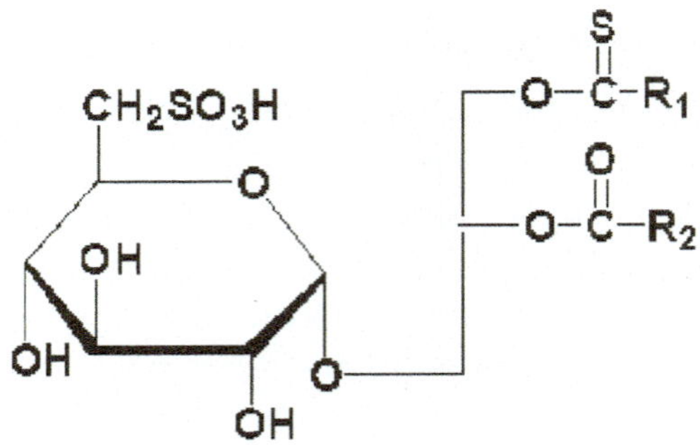

Thionsulfolipid

Thionsulfolipid producing cyanobacteria: *Thermosynechococcus elongatus* BP-1 (= *Synechococcus elongatus* BP-1)

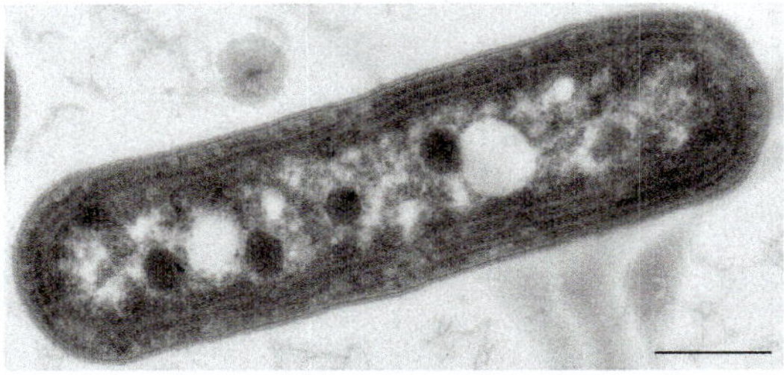

Thermosynechococcus elongatus
Image credit: Raul Gonzalez and Cheryl Kerfeld , Wikipedia

Global distribution of thionsulfolipid producing species: Japan

Habitat of thionsulfolipid producing species: Hot springs; it is a heat loving picocyanobacterium. It negatively and positively affected the coexisting filamentous cyanobacteria.

Hazards: although no health hazards have been reported so far due to this toxin, the crude and partially purified extracts of this toxin were found to be toxigenic to early life stages of aquatic invertebrates. Further this toxin reduced the condition of vertebrates and caused their behavioural disorders. Jakubowska aandaSzel g-Wasielewska, (2015) reported an LD50 value of 20 mg kg/l in a test conducted on minnows (small freshwater fish).

b) Microviridin J: It is a cyanopeptide toxin consisting of acetylated chain of 13 amino acids and is produced by the cyanobacterium strain *Microcystis* strain UWOCC MRC. Its entire toxicological profile is not known, albeit it is a protease inhibitor (Rohrlack et al., 2004) .

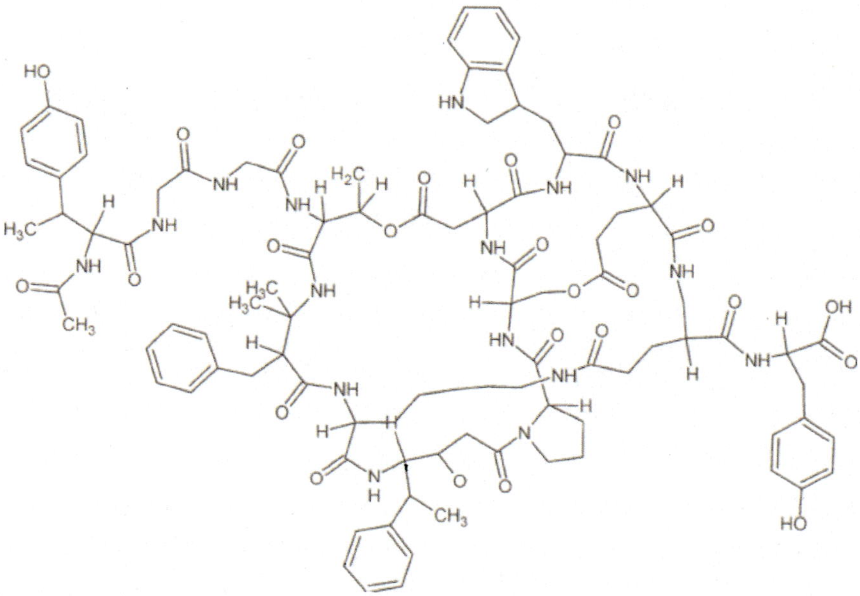

Microviridin J

2.1.1.2. Harmful Freshwater Cyanobacterial Blooms: Formation and Prevention

While the algal communities form an important component of healthy freshwater ecosystems such as ponds, lakes and rivers ,increased concentrations of nutrients such as nitrogen and phosphorus and temperature lead to "Freshwater harmful algal blooms (FHABs) ". The toxins produced by the latter are hazardous to freshwater ecosystems, animals and even to human health and the nation's economy. In recent years, the frequency and geographic distribution of these FHABs have been found to be increasing.

Toxic blue-green algal bloom in a reservoir

Factors Contributing to HAB Formation

Factors like nutrient concentrations, pH, water temperature, amount of light, and water circulation are mainly responsible for the formation of FHABs. Among these factors. Nutrient enrichment is mainly considered as an important one in HAB formation. When high concentrations of nitrogen and phosphorus enter the freshwater bodies , they not only induce the algal growth but also deplete the dissolved oxygen, reduce transparency and changes in biological community. This entire process is called eutrophication. The entry of nutrients to the freshwater systems may be either natural or by anthropogenic activities with point and nonpoint sources. While point sources involve industrial and municipal wastewater discharges and concentrated animal feeding operations, non-point sources include failing septic systems, manure runoff from agricultural lands and animal feeding operations (https://crsreports.congress.gov/product/pdf/R/R44871,2020).

Impacts of Harmful Cyanobacterial Blooms

FHABs may discolor the water column and the colorations of such blooms may be dull to bright green, yellow, brown or red due to the nature of algal species. Further, the texture of the blooms may be slicks, scums, mats of foam depending on the density of the algal species. Though several classes of algae are responsible in the development of FHABs in freshwater ecosystems, the cyanobacteria (blue-green algae) have been reported to be mainly responsible for the most frequent and harmful blooms which pose threat to human and animal components. The cyanotoxins produced by the harmful cyanobacterial blooms may pose a threat to the health of freshwater ecosystem and its plants and animals; and to humans . Water quality deterioration associated with depletion of oxygen is largely due to the formation of dead zones resulting with death and decomposition of algae. Humans are largely affected when they are exposed to cyanotoxins by drinking harmed water or eating finfish, or shellfish; recreating or swimming in waters with cyanotoxin ; and inhaling aerosolized cyanotoxins of these blooms. The affected humans have been reported to develop hepatic, dermatologic, neurologic and respiratory problems which may be fatal too. The types of cyanotoxins of harmful cyanobacterial blooms and their impacts on human beings have already been dealt with

Management practices for the mitigation of FHABs in freshwater bodies

Mitigation of FHABs include i) bloom prevention, ii) bloom reduction and prevention of future blooms, and iii) bloom reduction with no prevention of future blooms (https://mywaterquality.ca.gov/habs/images/algae_mitigation_technique_seletion_process_for_lakes.jpg) . Better preventive measures would not only help reduce FHABs but would also reduce their toll on ecosystems ,human health and national economies. The first and foremost preventive measure is to adopt better regulations to mitigate nutrient pollution (Español , https://www.nrdc.org/stories/freshwater-harmful-algal-blooms-101,2019).In order to achieve this, the following measures are suggested.

- Proper maintenance of septic systems
- Proper disposal of pet wastes
- Control of soil erosion and runoff
- Control the use of fertilizers and pesticides

Research gaps in FHABs

Intensive research on the following is the need of the hour in the mitigation of FHABs .

- Role of excess nutrients in the occurrence of FHABs
- Understanding less-common FHAB species which produce toxins
- Development of standardized methods for algal toxins
- Predicting the timing, species composition, and toxicity of FHABs
- Strengthening the existing monitoring programs
- Development of effective methods for mitigating FHABs
- Understanding of the environmental and economic impacts of HABs
- Understanding of the impacts of FHAB exposure on human health
- Improvements in communication tools and public outreach

Human health effects of freshwater cyanobacterial toxins

The human health impacts from the various freshwater cyanobacterial toxins are given in the following table.

Toxin	Species	Impacts on human health
Microcystins	*Synechocystis aqualitis, Synechococcus,* sp., *Cyanobium rubescens, Aphanocapsa cumulus, Merismopedia* sp., *Romeria carauru, Microcystis aeruginosa, Anabaena* sp., , *Planktothrix* sp., *Nostoc* sp., *Hapalosiphon* sp., *Anabaenopsis* sp.	Gastrointestinal, liver inflammation, and hemorrhage and liver failure leading to death, pneumonia, dermatitis; Tumor promoter, liver failure leading to death
Nodularins	*Nodularia* sp., *Merismopedia tenuissima, Aphanothece stratus, Aphanothece parallelliformis, Aphanocapsa delicatissima, Aphanocapsa incerta, Cyanodictyon planctonicum, Cyanodictyon imperfectum, Synechococcus* sp.,	Gastrointestinal, liver inflammation, and hemorrhage and liver failure leading to death, pneumonia, dermatitis; Tumor promoter, liver failure leading to death

Toxin	Species	Impacts on human health
Anatoxin-a	*Anabaena flos-aquae, Planktothrix* sp., *Aphanizomenon* sp.	Tingling, burning, numbness, drowsiness, incoherent speech, respiratory paralysis leading to death; Cardiac arrhythmia leading to death
Anatoxin-a(s)	*Anabaena* sp.	Tingling, burning, numbness, drowsiness, incoherent speech, respiratory paralysis leading to death; Cardiac arrhythmia leading to death
Aplysiatoxin	*Lyngbya* sp., *Planktothrix* sp., *Oscillatoria nigroviridis, Nostoc muscorum*	Dermatitis
Cylindrosper-mopsin	*Cylindrospermopsis* sp., *Aphanizomenon* sp., *Umezakia* sp.	Gastrointestinal, liver inflammation and hemorrhage, pneumonia, dermatitis; Malaise, anorexia, liver failure leading to death
Lyngbyatoxins	*Lyngbya* sp.	Dermatitis; Skin tumors
Saxitoxins	*Anabaena* sp, *Aphanizomenon* sp., *Lyngbya* sp., *Cylindrospermopsis* sp.	Tingling, burning, numbness, drowsiness, incoherent speech, respiratory paralysis leading to death
Lipopolysac-charide	*Synechocystis* sp., *Synechococcus* sp., *Aphanocapsa* sp.,	Gastrointestinal, dermatitis
aBeta -N-me-thylami-no-l-alanine (BMAA)	*Microcystis* sp., *Synechocystis* sp., *Synechococcus* sp., *Nostoc* sp., *Planktothrix agardhii, Anabaena variabilis, Cylindrospermopsis raciborskii,*	Neurodegenerative diseases; Amyotrophic Lateral Sclerosis (ALS) /Parkinsonism-dementia complexa(PDC)

Source: Lopez,et al., 2008; Vyasa and Weiss, 2009

2.1.2. Green algae (Chlorophyta; Chlorophyceae)

Toxins: In the freshwater green algae no toxins have been identified so far. However, these algal blooms may be harmful to the biota of such water bodies and humans directly or indirectly.

Harmful green algal species: *Botryococcus* spp., *Cholorococcus* spp. , *Dictyosphaerium* spp., *Pandorina indica, Pandorina morum, Scenedesmus* spp., *Sphaerocystis* spp., and *Volvox aureus*

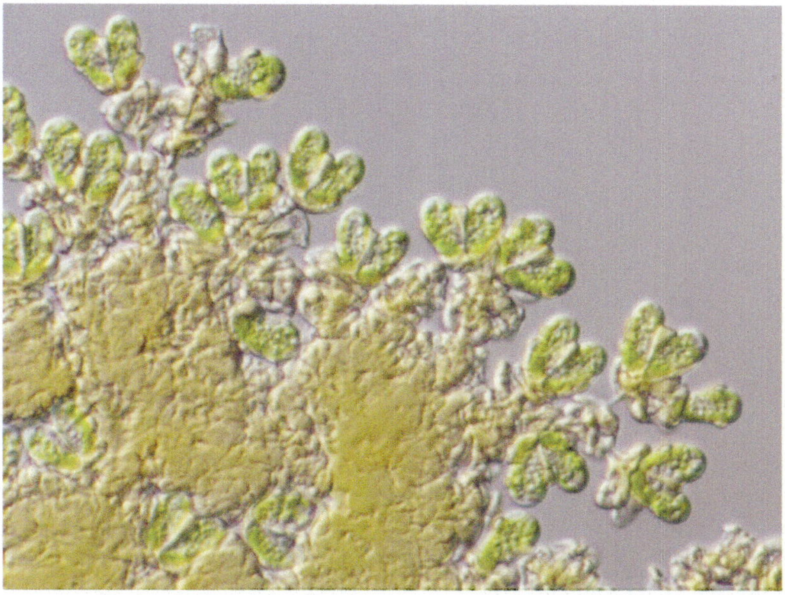

Botryococcus **sp.**
Image credit: ja:User:NEON , Wikipedia

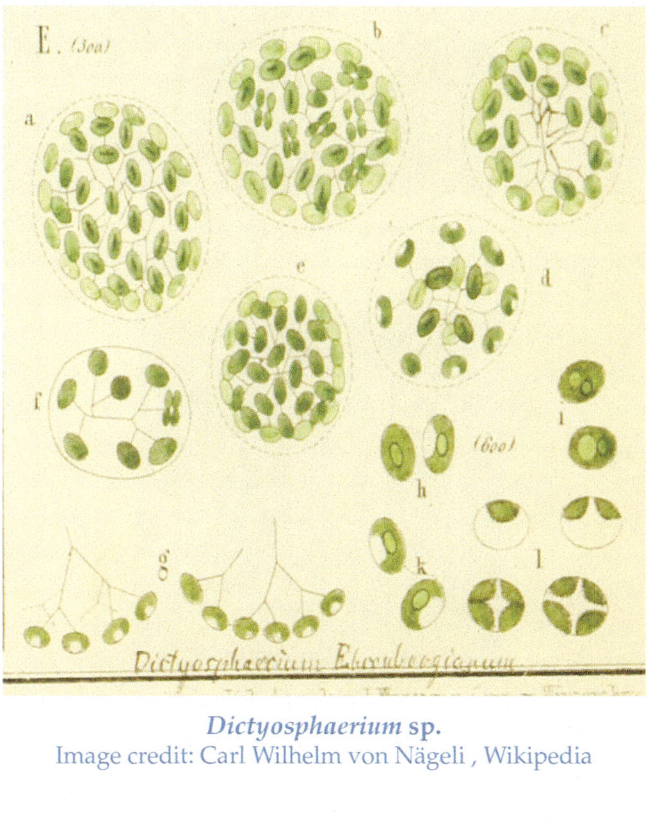

Dictyosphaerium **sp.**
Image credit: Carl Wilhelm von Nägeli , Wikipedia

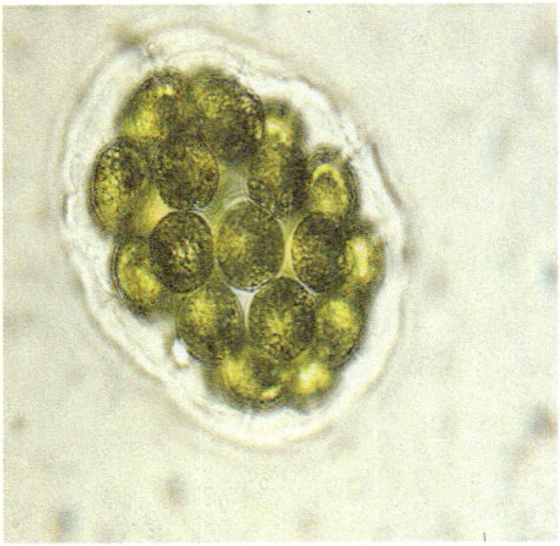

Pandorina **sp.**
Image credit: Wikispecies

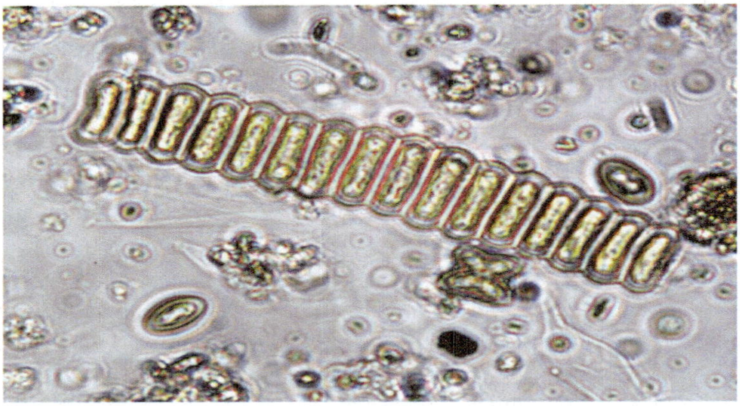

Scenedesmus sp.
Image credit: Environmental Protection Agency , Wikipedia

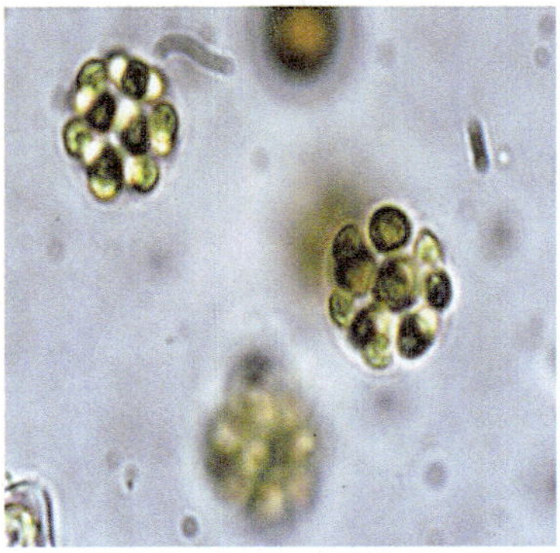

***Sphaerocystis* sp.**
Image credit: Environmental Protection Agency, Wikipedia

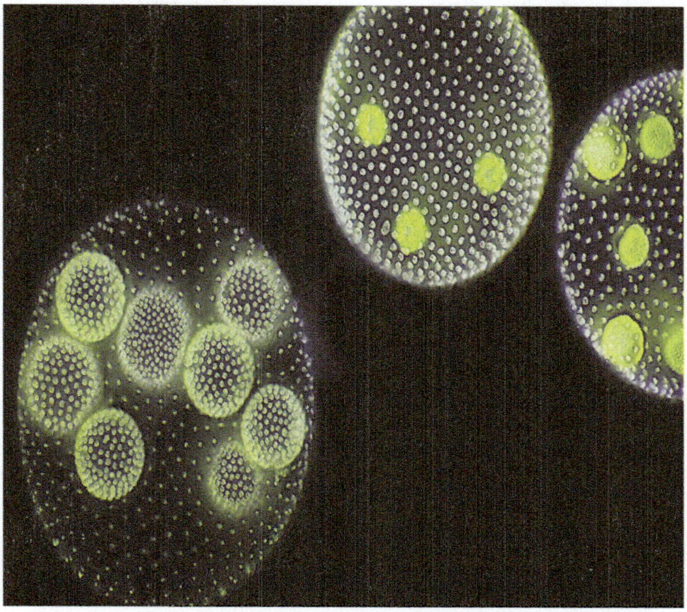

Volvox sp.
Image credit: Frank Fox , Wikipedia

Environmental/ harmful effects: The aforesaid chlorophyte genera are known for the formation of distinctive blooms and discoloration of water bodies. On rare occasions these organisms may proliferate and water quality problems such as hypoxia, odor and taste. Species of *Botryococcus* may form highly buoyant surface-dwelling aggregates which can accumulate scums and affect the world-class trout fishery of the lakes (Paerl et al., 2001). Similarly *Pandorina morum* toxin has been reported to inhibit the growth of algae, bacteria and higher plants (Patterson and Harris, 1983)

2.1.3. Euglena (Euglenophyta; Euglenophyceae)

Toxins: Euglenophycin is an alkaloid

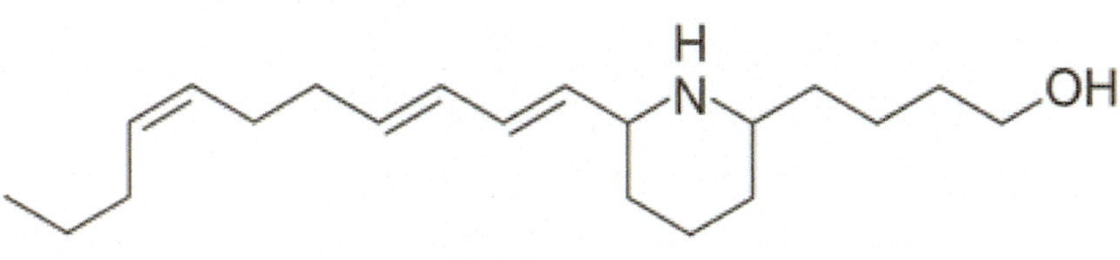

Euglenophycin

Euglenophycin producing species : *Euglenophycin is produced by aEuglena sociabilis, Euglena stellata, Euglenaria clavataa, Euglenaria anabaena, Strombomonas borystehnienisa, Trachelomonas ellipsoidalis,* and *Lepocinclis acusa*

***Eugelna* sp.**
Image credit: Wikipedia

***Trachelomonas* sp.**
Image credit: ja:User:NEON , Wikipedia

***Lepocinclis* sp.**
Image credit: Picturepest , Wikipedia

Global distribution of Euglenophycin producing species: Species of *Euglena* are found worldwide

Habitat of Euglenophycin producing species: Species of *Euglena* are found in shallow, calm, and eutrophic freshwater systems such as ponds rich in organic matter. Some species may form "green or red blooms" in ponds or lakes.

Environmenta/ Harmful effects of euglenophycin : Species producing >5 g/cell euglenophycin included single species of the genera *Lepocinclus, Trachelomonas* and *Stombomonas*; a two species of *Euglenaria*; and three species of *Euglenaa* (including E. *sanguinea*). Two taxa, *Euglena socialis* and *E. sanguinea*, were however found to accumulate euglenophycin at levels 100- and 1000-fold higher than other Euglenacea. *Euglena sanguinea* is the only known species of euglenids forming toxic blooms which are associated with discoloration of water and mortalities of fish. Consumption of affected fish may cause health problems (Zimba et al.a2017).

2.1.4. Raphidophytes (Ochrophyta ; Raphidophyceae)

Species of Raphidophyceae such as *Gonyostomum semen, Merotrichia* sp. , and *Vacuolaria* have been reported to produce ichthyotoxins. Among them *Gonyostomum semen* is a nuisant algal species causing fish kills (Leblond et al., 2013)

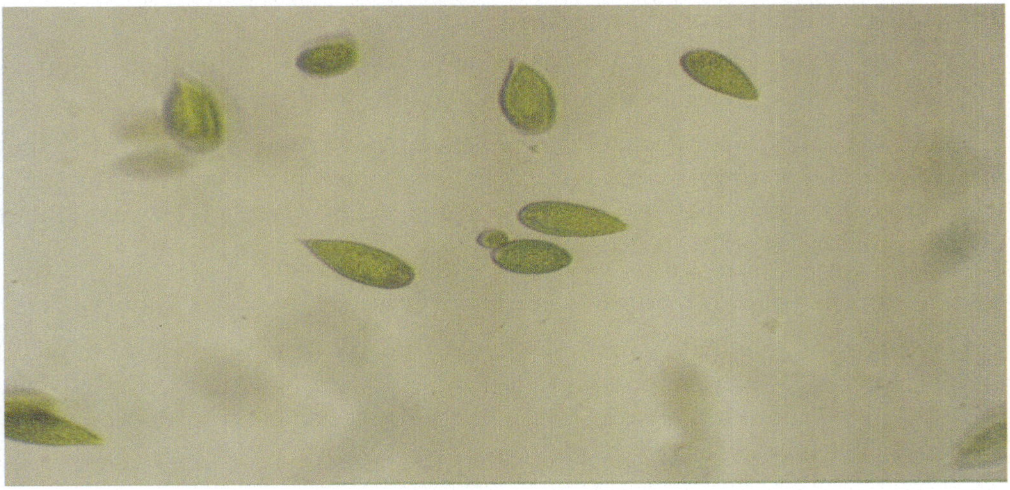

***Gonyostomum* sp.**
Image credit: Krengefors , Wikipedia

2.1.5. Cryptomonads (Cryptophyta : Cryptophyceae)

Toxins: Cryptomonads (Rocky mountain lake algae) are not known to produce toxins.

Cryptomonad species causing environmental pollution: – *Rhodomonas, Cryptomonas* and *Goniomonas*

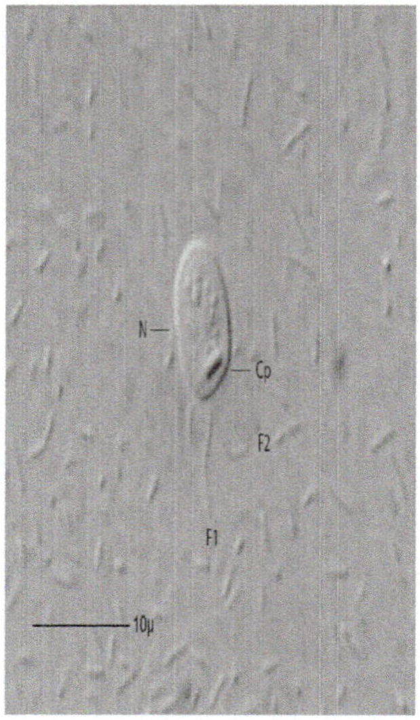

Goniomonasa **sp.**
Image credit: Picturepest, Wikimedia Commons

Global distribution of cryptomonads: Worldwide

Habitat of cryptomonads: While Goniomonas thrives in all aquatic habitats, *Rhodomonas* and *Cryptomonas* are found only in freshwater habitats

Hazards of cryptomonads: Cryptophytes are not known to be toxic to either human or resident animals and plants. However, these organisms in association with dinoflagellates cause "red tides", localized hypoxia and adverse conditions in their habitats. (Paerl et al., 2001)

2.1.6. Diatoms (Bacillariophyta; Bcillariophyceae)

Neurotoxins: β-N-methylamino-L-alanine (BMAA) (a non-protein amino acid) and its isomers 2,4-diaminobutyric acid (2,4-DAB), and N-(2-aminoethyl)glycine (AEG). Among these isomers, 2,4-DAB, like BMAA, has neurotoxic properties (Violi et al., 2019)

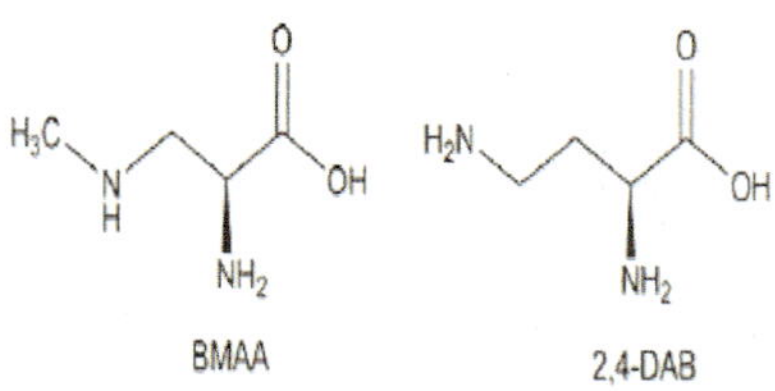

2,4-DAB

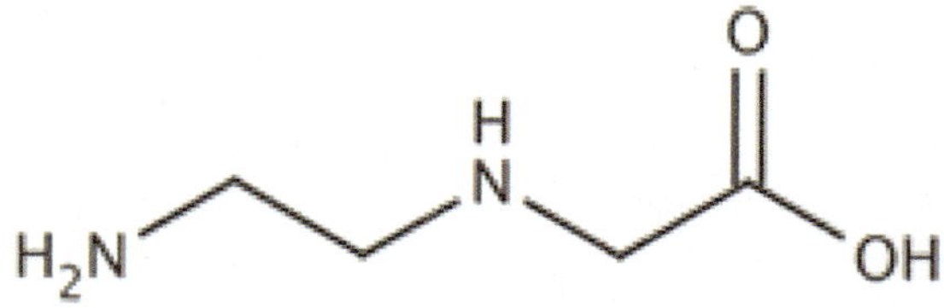

AEG

BMAA producing diatoms: *Thalassiosira, Aulacoseira, Cyclotella, Fragilaria, Navicula, Tabellaria* and *Didymosphenia*

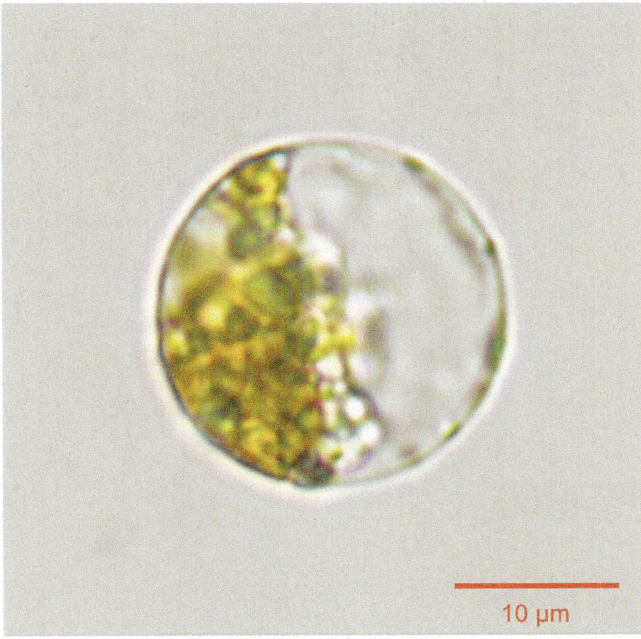

Thalassiosira pseudonana
Image credit: Oregon State University, Wikipedia

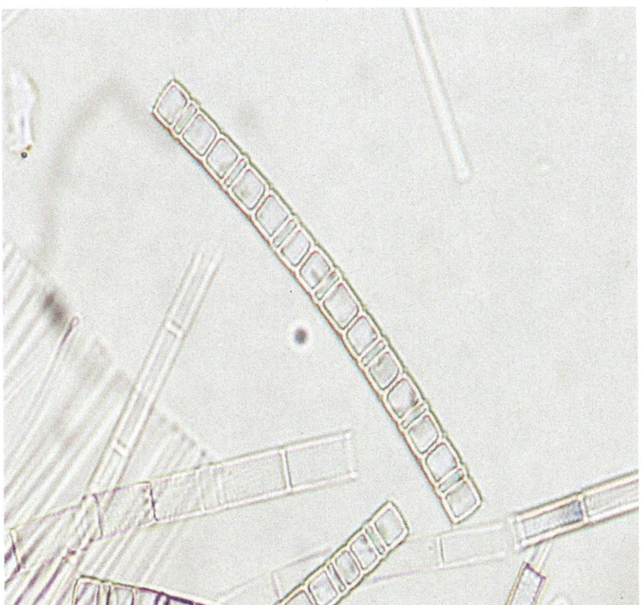

***Aulacoseira* sp.**
Image credit: Kristian Peters , Wikipedia

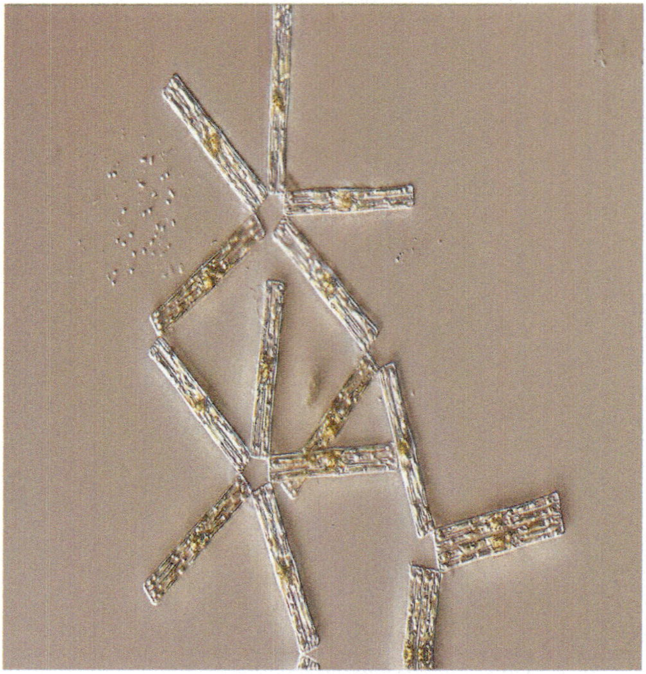

***Tabellaria* sp.**
Image credit: Kairi Maileht , Wikipedia
Image credit: Wikimedia Commons

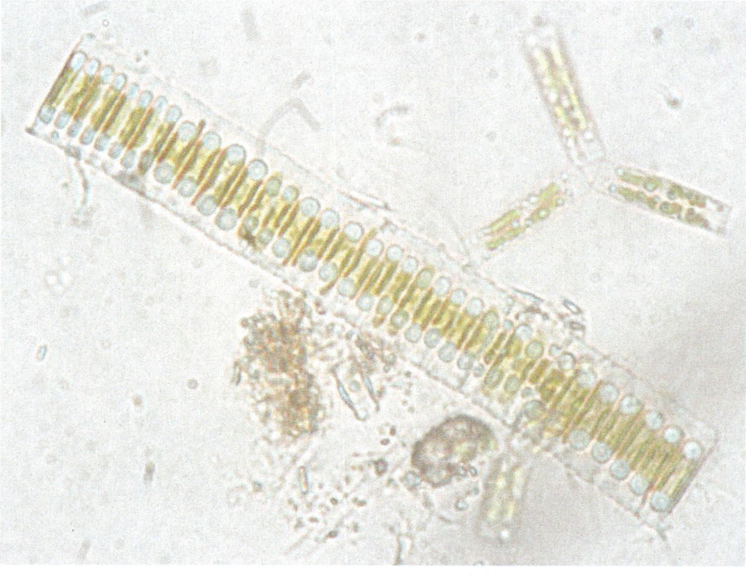

Fragilaria
Image credit: Wikimedia Commons

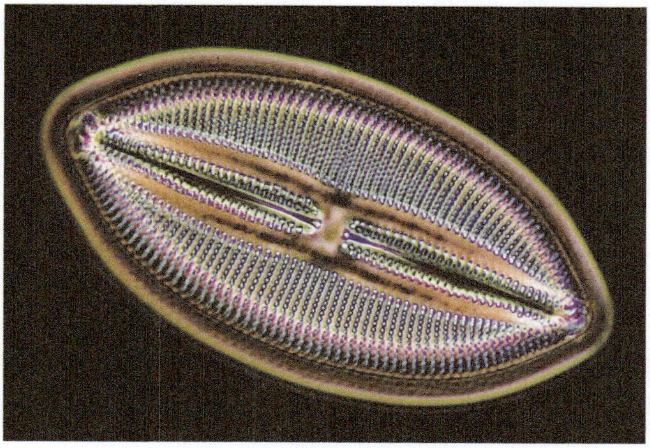

***Navicula* sp.**
Image credit: Massimo brizzi , Wikimedia

***Didymosphenia* sp.**
Image credit: Djpmapleferryman, Wikimedia Commons

BMAA producing diatoms

Global distribution of BMAA producing species: World-wide distribution

Habitat of BMAA producing species: Freshwater habitats

Health hazards of BMAA: BMAA has been implicated as a risk factor for motor neurone disease. BMAA and its isomers viz. 2,4-diaminobutyric acid (2,4-DAB), and N-(2-aminoethyl)glycine (AEG) have been isolated from diatom genera viz. *Thalassiosira, Aulacoseira, Cyclotella, Fragilaria, Navicula* and *Tabellaria* . Among these components BMAA and 2,4-DAB showed neurotoxic properties. It was also observed that *Fragilaria* did not have detectible levels of BMAA; *Cyclotella* did not have detectable levels of AEG; and 2,4-DAB was detected in all the diatom genera. The highest BMAA, 2,4-DAB and AEG concentrations were detected in *Navicula* at 369.64 ng/g, 4678.75 ng/g and 1328.11 ng/g per dry weight (DW) respectively. The lowest BMAA concentration per dry weight was detected in *Tabellaria* at 19.99 ng/g . The lowest 2,4-DAB concentration per dry weight was detected in *Aulacoseira* at 103 ng/g. The lowest AEG concentration per dry weight was detected in *Aulacoseira* at 49.55 ng/g. It is also reported that the neurotoxins of the aforesaid diatom genera may pose a health risk to people living nearby freshwater sources via drinking water contamination, aerosolisation and/or the consumption of food source (including contaminated fish) (Violi et al., 2010)

Didymosphenia geminata has yielded BMAA, a neurotoxin at a concentration of 0.17 – 0.28 µg/ g DW. The toxicology of this species on humans is yet be known (file:///C:/Users/rsant/Downloads/160727_Reveillon_revisedarticle_archimer.pdf),

Spaulding and Elwell (2007) reported on the harmful effects of the diatom, *Didymosphenia geminata.* During bloom conditions, this species produces copious amounts of extracellular stalk material which forms thick benthic mats which altered the species diversity and population sizes of the algae, invertebrates, and fish. Further the extensive mats of *D. geminata* also impacted the ecological processes such as ecosystem metabolism and nutrient cycling. Furthermore, *D. geminata* was predicted to possess deleterious effects on native fish populations. Nuisance growths of this species had the potential to affect the fisheries through food web interactions with macroinvertebrates.

2.1.7. Prymnesiophytes (Haptophyta; Class: Prymnesiophyceae)

Toxins: The toxic substances of *Prymnesium parvum* have been reported to include lipopolysaccharides (hemolysins), a galactoglycerolipid, polene polyethers, cyclo-amines, dimethylsulfonio-propionate, polyunsaturated fatty acids, and fatty acid amides . As examples, the complex polyketides prymnesin-1 and -2 have been isolated from this species and these toxins have been reported to possess ichthyotoxic, neurotoxic, hemolytic, and cytotoxic properties (Roelke and Manning,a2018). The other reported toxin of this species is prymnesin B-1 which is a polyether phycotoxin.

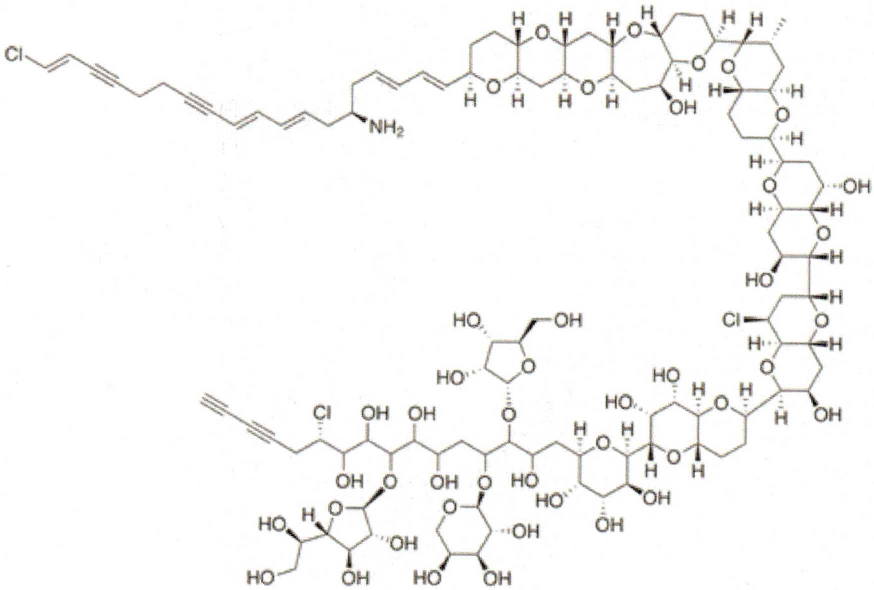

Prymnesin-1

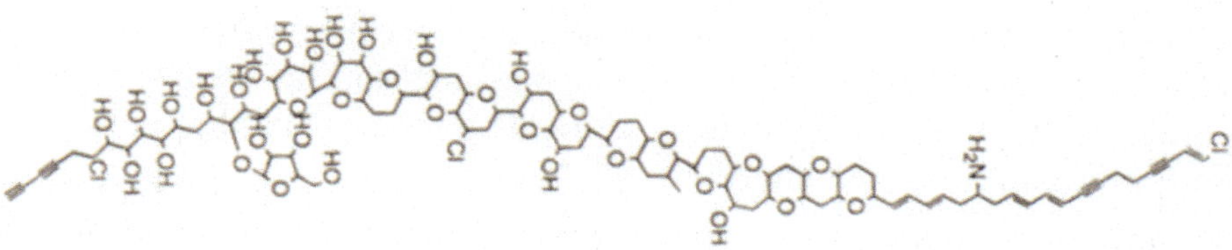

Prymnesin-2

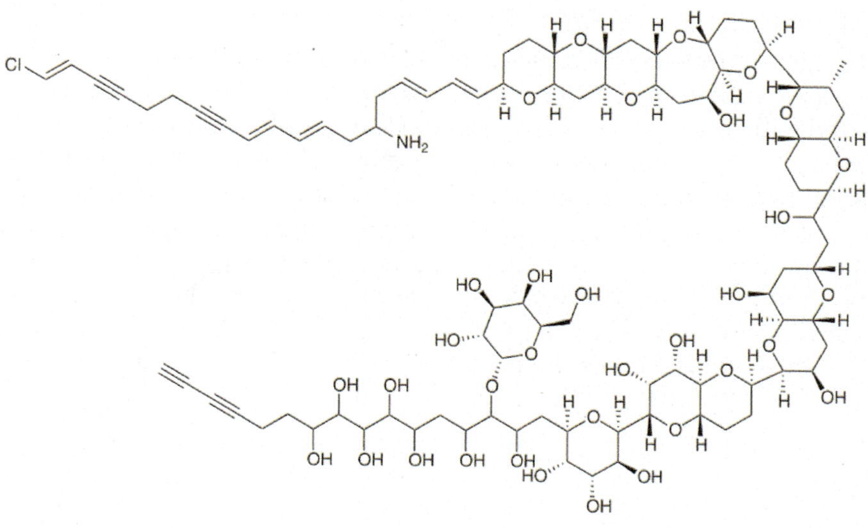

Prymnesin-B1a

Prymnesin-producing species : *Prymnesium parvum* (= *Prymnesium parvum* f. *patelliferum*) (Prymnesiophyceae ; Prymnesiaceae)

Prymnesium parvum

Image credit: Daniel Killerup Svenssen, Sofie Bjørnholt Binzer, Nikola Medi , Per Juel Hansen, Thomas Ostenfeld Larsen, Elisabeth Varga, Wikimedia Commons

Global distribution of Prymnesin-producing species: *Prymnesium parvuma* is found worldwide. While the harmful blooms of this species were recorded ed in the eastern hemisphere since the early 1900s, this species invaded in all southern regions of the USA and some northern regions.

Habitat of Prymnesin-producing species: This species ais found in brackish inland and coastal aquatic environments. Its freshwater habitat include rivers, lakes and ponds with a relatively high mineral content. This is an euryhaline organism tolerates salinities ranging from 1 to > 35 ppt (Roelke and Manning, 2018).

Ecological hazards of Prymnesins: The blooms of *Prymnesium parvum* are often monospecific with very high densities which make the golden colored and are often associated with large fish kills. These blooms are therefore commonly referred to as "golden alga" or "golden algae blooms". Further, these blooms may also be large, stretching along watersheds hundreds of kilometers (Roelke and Manning,a2018). The toxins of this species are mainly ichthyotoxic and are largely responsible for devastating fish kills causing ecological and economic damage. Blooms of this species often result in fish kills in 100,000s of individuals and costing local economies in the tens of millions of dollars (USD). In Texas , the losses to local economies during 2001 winter fish kills were reported to exceed $18 million (https://www.cabi.org/isc/datasheet/121720).

Human health hazards: The toxin prymnesin has been reported to possess potent neurological toxicity i.e. it interferes with the re-uptake of neurotransmitters at the post-neuronal synapse. Further, prymnesin inhibits acetylcholine and interfere with calcium signaling. It is also reported that an intravenous injection of prymnesins stopped the heart muscle by rapid depolarization and by altering the membrane permeability to calcium ions. In another research report, it is said that an injection of prymnesin evoked muscle contractions, which was followed by the reversal of the acetylcholine-induced contractions and paralysis . An intravenous injection of prymnesins also resulted in a sudden drop in blood pressure and respiratory arrest . Prymnesins are also known for their potent erythrolytic capacity and prymnesins 1 and 2 have been reported to be 5,000 times more hemolytic (Manning,2010).

2.1.8. Dinoflagellates (Dinophyceae; Peridiniaceae)

Toxins: Glenodinine which is an alkaloid with ichthyotoxic properties (Hashimoto et al., 1968).

Toxin producing dinoflagellates : *Naiadiniuma polonicuma (= Peridinium polonicum); Peridinium aciculiferum*

Global distribution of toxin-producing freshwater dinoflagellate species: It is a acosmopolitan freshwater species and is common in European lakes

Habitat of toxin-producing freshwater dinoflagellate species: This species occurs in all freshwater habitats

Hazards of toxin-producing freshwater dinoflagellate species: The ichthyotoxin glenodinine of this species has caused mortalities of fish (Hashimoto et al., 1968).

Toxin--producing dinoflagellate species : *Peridinium aciculiferum*

Peridinium aciculiferum

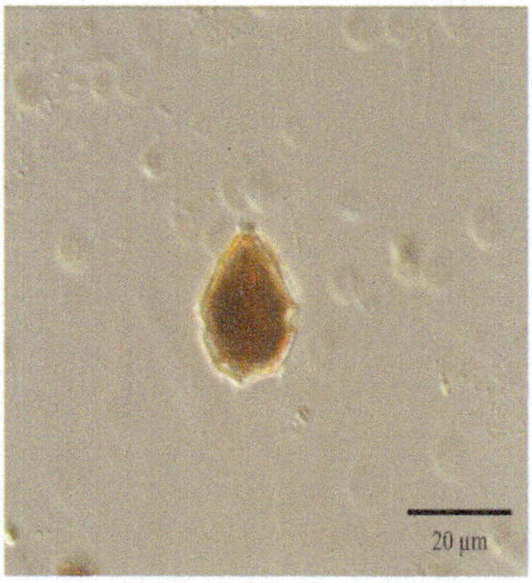

Image credit: Nordic Microalgae

Global distribution of *Peridinium aciculiferum*: Temperate lakes such as Swedish lakes ; and Ceylon freshwater bodies

Habitat of *Peridinium aciculiferum*: It forms blooms underneath the ice

Human health hazards of *Peridinium aciculiferum*: Yatigammana et al.(2011) reported on the occurrence of a dysenteric epidemic condition in a community of people who consumed water from the reservoirs contaminated by the blooms of *Peridinium aciculiferum*. The symptoms associated with this condition were diarrhoea, vomiting and headache

Ecological hazards of *Peridinium aciculiferum*: This species forms blooms during winter and it has been reported to be potently ichthyotoxic. The unidentified toxin of this species also possesses hemolytic activity (Rengefors and Legrand, 2007).

Paerl et al. (2001) reported on the formation of blooms of this species with high densities (> 100,000 cells/ml) and associated water quality conditions economic damage viz. foul odors and hypoxia ; and loss of recreational value and economic loss.

2.2. Toxic Freshwater Macroalgae

Cladophora glomerata

Image credit: W.carter, Wikipedia

Global distribution of *Cladophora glomerata*: It is the most widely distributed macroalga throughout the world's freshwater ecosystems including North America, Europe, Asia, Africa, Australia and New Zealand.

Habitat of *Cladophora glomerata* : It is found in a variety of marine and freshwater bodies as either attached or drifting form. It is known to provide habitat and food for numerous organisms of the same habitat.

Ecological hazards of *Cladophora glomerata*: This species is considered a nuisance in recreational bodies of water and is often causing localized hypoxia.It has been reported that the overgrowth of this species is often associated with the rise of invasiveazebra mussels in the Great Lakes of North America (https://www.britannica.com/science/Cladophora#ref1278905). The zebra mussels (*Dreissena polymorpha*) which was originally native to the lakes of southern Russia and Ukraine have caused biofouling albeit they are insignificant in size.

Human health hazards of *Cladophora glomerata*: Ishii et al. (2006) reported that this macroalga harbored Shiga toxin (Stxs) -producinga pathogenic bacteria such as *Escherichia coli, Salmonella, Shigella,* and *Campylobacter* in the Lake Michigan. Humans infected with Stx-producing *Escherichia coli* showed the most serious manifestation of the disease viz. hemolytic uremic syndrome (HUS) which may occur when the small blood vessels in kidneys become damaged. This damage can cause clots in the vessels and associated clogging in the filtering system of the kidneys leading to kidney failure. The Stxs (also known as Vero toxins) have also been reported to inhibit protein synthesis in sensitive eukaryotic cells.

Capric acid

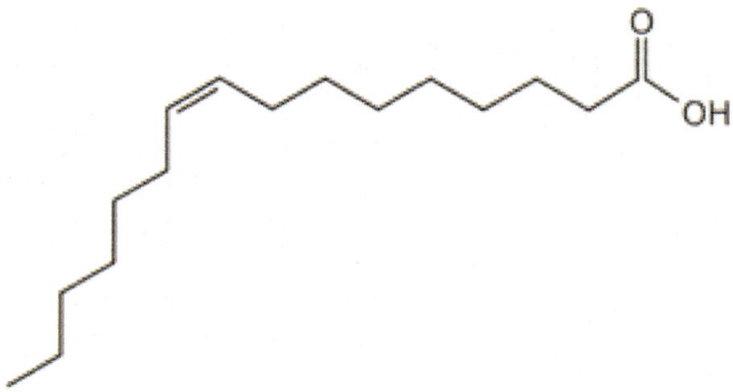

Palmitoleic acid

CHAPTER 3

POISONOUS FRESHWATER PLANTS

Hygrophila balsamicaa

***Hygrophila* sp.**
Image credit: Wikipedia

Family: Acanthaceae

Common name (s) : Giant aqua rose, aromatic swampweed

Global distribution: It is native to India and Sri Lanka

Ecology: It is a marsh dwelling plant and is well-suited for tanks with lots of light and a high CO_2 content

Description: It is an erect herb and is growing up to 50 cm tall. Stem is well-branched and hairless. Leaves are opposite; obovate-elliptic to lance-shaped and are with finer pinnate leaf structure. Flowers which are borne in verticullate cymes are 1.2 cm across and are blue. It is a fast grower and if there is less intensive light and a lack or a deficiency of CO_2, it will not attain its characteristically beautifuly bright green hue.

Harmful organs and harm: This plant's sap is poisonous to fish (https://www.aquasabi.com/Hygrophila-balsamica-pot)

Acer rubrum

Image credit: Famartin, Wikipedia

Family: Aceraceae

Common name: Red maple

Global distribution: Eastern and Central North America

Ecology: It is commonly found in swampy areas, banks of streams, wet bottomland and river flood plains.

Description: It grows to a maximum height of about 20 m and is with a rounded to oval crown. Bloom is red or yellow; and leaves are medium to dark green above and gray green below. Flowers are monoecious (male or female) and they appear in late winter to early spring (March-April) before the appearance of leaves

Harmful effects : Leaves of red maple, especially when dead are extremelyatoxic to animals. Its toxin is anaoxidanta as it damagesared blood cells, causing acuteaoxidative hemolysisawhich inhibits the transport ofaoxygen to the tissues . This leads to the production ofamethemoglobin, which can further damage the kidneys. Ingestion of 700ag of leaves is toxic and 1.4kg is lethal. Symptoms of illness which occur within a day or two after ingestion include depression, lethargy, increased rate of breathing, increased heart rate,ajaundice, dark brown urine, and coma, and finally death in affected animals (https://en.wikipedia.org/wiki/Acer_rubrum)

Amaranthus retroflexus

Image credit: Lynk media , Wikipedia

Family: Amaranthaceae

Common name (s) : Redroot pigweed, Careless Weeda

Global distribution: Central and eastern USA; south-eastern Canada; and north-eastern Mexico

Ecology : It is commonly seen in river banks and bottoms

Description: It is an annual plant with a height varying from 30 to 180 cm. Leaves are alternate and diamond shaped. Male and female flowers are separate , small and they form long clusters

Harmful effects: These plants have been reported to be the chief cause of hay fever in humans. The renal toxicity of this poisonous plant is due to its oxalic acid which is at 2–5%. After the consumption of this plant for several days, renal disease develops and it is followed by diarrhea., trembling, paralysis, coma, and death in a period of about 2 days (Allison, https://pubs.nmsu.edu/_circulars/CR678/)

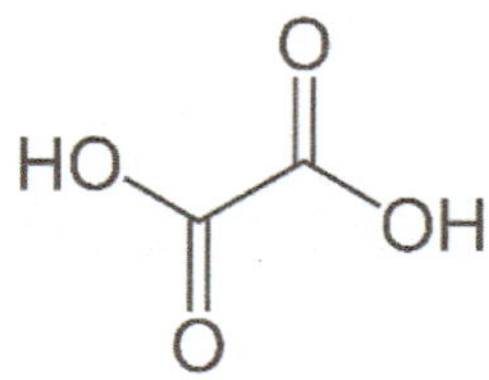

Oxalic acid

Crinum americanum

Image credit: Vincent P. Lucas , Wikipedia

Family: Amaryllidaceae

Common name (s) : American crinum lily, swamp lily, Florida swamp-lily, seven sisters

Global distribution: It is native to the southern United States from Florida to Texas

Ecology: It occurs in gardens, ponds ,streams; marshes and swamps and along river banks

Description/ Biology: This perennial plant spreads quickly and produces white flowers with six petals. Due to its aggressive tendencies, this water plant can form heavy clusters that interfere with oxygen and sunlight levels in standing water features

Harmful effects: All parts of this plant including seeds are poisonous and may cause stomach upset including abdominal pain, nausea and vomiting and diarrhea if ingested. (https://www.ehow.com/list_6628387_poisonous-water-plants.html ; https://davesgarden.com/guides/pf/go/1172/#b)

Rhus vernix

Image credit: Wiktionary

Family: Anacardiaceae

Common name (s) : Poison sumac and swamp-sumac

Global distribution: It is found distributed in southern United States

Ecology : It is also seen in stream borders.

Description: It is a shrub or small tree growing to a maximum height of 5 m. Leaves are alternate, and are pinnately compound. There are 7-13 elliptic and entire leaflets. Flowers are seen in panicles in the axils of lower leaves.

Harmful effects: The toxicity of this plant is due to a phenolic compound called urushiol which is seen in all parts of the plant. Animals have been reported to be rarely affected by this plant. As its toxin is a skin and mucous membrane irritant,, humans who are sensitive are said to be affected by this plant on contact (https://carteret.ces.ncsu.edu/wp-content/uploads/2013/05/Poisonour-Plants-of-the-Southern-United-States.pdf?fwd=no). It is also reported that the inhalation of the smoke by burning the plant may cause lung damage with extreme pain and possibly fatal respiratory difficulty.

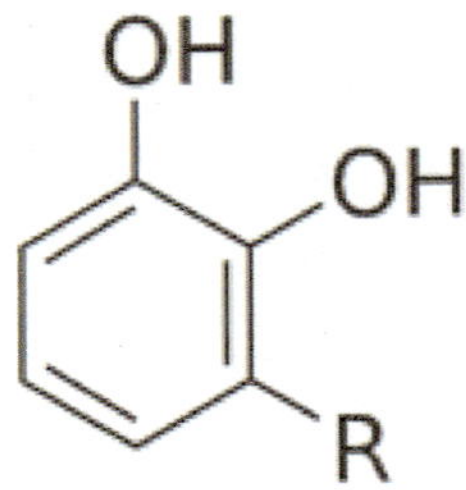

Urushiol

Cicuta douglasii

Image credit; Wikipedia

Family: Apiaceae

Common name (s) : Western water hemlock

Global distribution: North America.

Ecology : This plant is seen along irrigation canals and streams

Description: This perennial plant grows to a maximum height of about 1 m. It has small, white flowers which seem to be umbrella like clusters.

Harmful effects: The toxic compound in this poisonous plant is is cicutoxin, which is a highly poisonous unsaturated alcohol found mainly in the tubers, but it may also be seen in in the leaves and stems especially during early growth. Only a small amount of this toxic compound is needed to produce poisoning in humans. Its cicutoxin has been reported to be a violent convulsant acting directly on the central nervous system. The symptoms associated with its unintentional consumption are nervousness, rapid breathing, violent convulsions and coma. Death may result in 15 min after a lethal dose is consumed (USDA Agricultural Research Service)

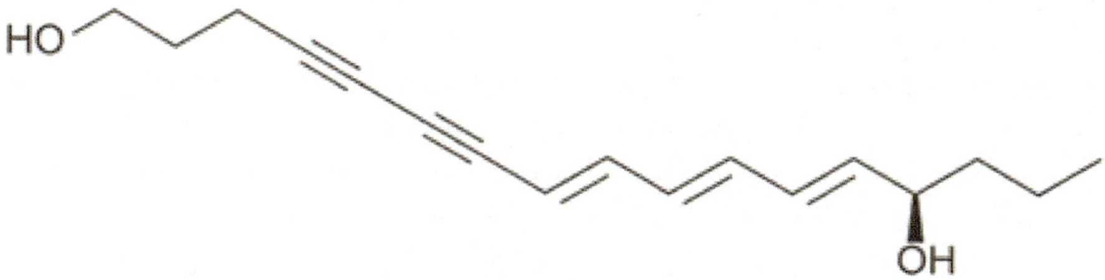

Cicutoxin

Cicuta maculata

Image credit: Williammehlhorn , Wikipedia

Family: Apiaceae

Common name (s) : Spotted cowbane, spotted water hemlock, spotted parsley

Global distribution: It is native to North America; from northern Canada to southern Mexico

Ecology : It is found in the moist soils of marsh edges, prairies and meadows. It can however, tolerate standing water for only short periods

Description: It is a rhizomatousa perennial herb and is producing a hollow erect stem which reaches a maximum height of 1.5 m. Its long leaves are made up of several lance-shaped, pointed, serrated leaflets. Each shiny leaflet is 2 to 10 cm long and the entire leaf may reach a total length of 40 cm. Inflorescence is of white flowers and it is a compound umbel with many clusters of flowers.

Harmful effects: This species is considered to be the most poisonous plant in North America. Its toxicity of this plant lies mainly in its green stem , leaves and roots. It roots are known to concentrate a poison known as 'cicutoxin' which can kill a cow, hence its name Cowbane. Symptoms associated with its toxicity include sweating, vomiting, violent salivating, kidney failure, irregular heartbeat and difficulty breathing. Nosowitz (2017) reported that its cicutoxin makes convulsions in affected persons. During these convulsions, every single muscle starts firing and contracting; the tongue is chewed into ribbons and the mouth cannot be opened as the jaw muscles are contracting about 20 times which lead a horrible death. It is also believed that the muscles of the affected persons may start contracting so hard and they can even dislocate bones. According to this author, the person may die in two hours after the unintentional consumption of this root. Further, its cicutoxin is an antagonist of gamma-aminobutyric acid (GABA) , which is the neurotransmitter of humans.

Conium maculatum

Image credit: Djtanng , Wikipedia

Family: Apiaceae

Common name (s) : Poison hemlock, wild hemlock

Global distribution: Western Asia , Europe and Northern Africa

Ecology : It is seen in open flood plains of rivers and streams

Description: It is a herbaceous biennial flowering plant growing to a maximum height of 3.6m. Stem is hollow, smooth and green. Leaves are 2-4 pinnate and triangular shaped. Flower is white and small with five petals.

Harmful effects: All parts of this plant are poisonous. The toxicity of this plant is due to its neurotoxin viz. piperidine alkaloids including coniine and γ-coniceine. Consuming even a little bit of any part of this plant may kill humans and the symptoms appear in 20 minutes after ingestion. The symptoms of poisoning include dilation of the pupils, trembling , paralysis of the central nervous system, slowing of the heartbeat, muscle paralysis, and death due to mainly respiratory failure (https://kingcounty.gov/services/environment/animals-and-plants/noxious-weeds/weed-identification/poison-hemlock.aspx). Severe poisoning of this plant has also been reported to cause rhabdomyolysis which may lead to acute renal failure. It is also believed that skin exposure to this plant may cause dermatitis (USDA Agricultural Research Service)

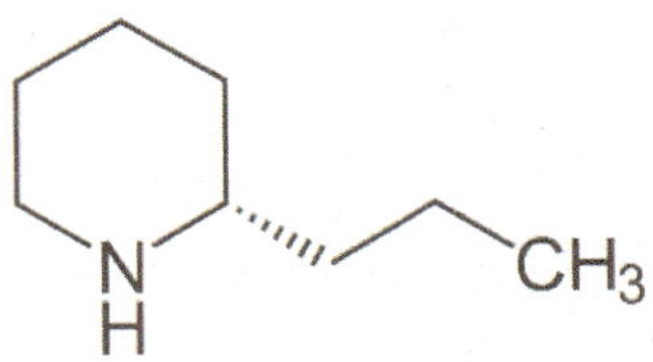

Coniine

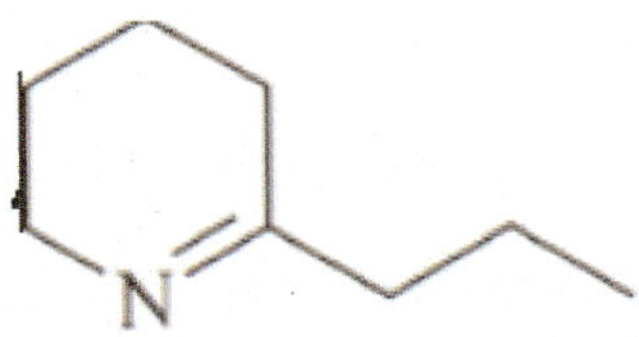

γ-coniceine

Oenanthe crocata

Image credit: Krzysztof Ziarnek, Kenraiz , Wikipedia

Family: Apiaceae

Common name (s) : Hemlock water-dropworta; dead man's fingers

Global distribution: It is native to western Asia, Europe and North Africa

Ecology : it is commonly seen along stream and river banks.

Description: This perennial plant grows up to about 150 cm. Leaves are spirally arranged. While the upper leaves (flat leaflets) are 1–2 pinnate, with a shorterapetiole, the lower leaves are pinnate in 3–4 times and are triangular with cuneate stalks. It has compound umbels.

Harmful effects: All parts of this plant have been reported to be extremely toxic to humans. The toxiccompound of this plant is oenanthotoxin, a polyunsaturated higher alcohol which is a strong convulsanta blocking the γ-aminobutyric acida(GABA) system in the human brain. Symptoms associated with its unintentional consumption include vomiting, lung collapse and haemorrhaging of the brain.

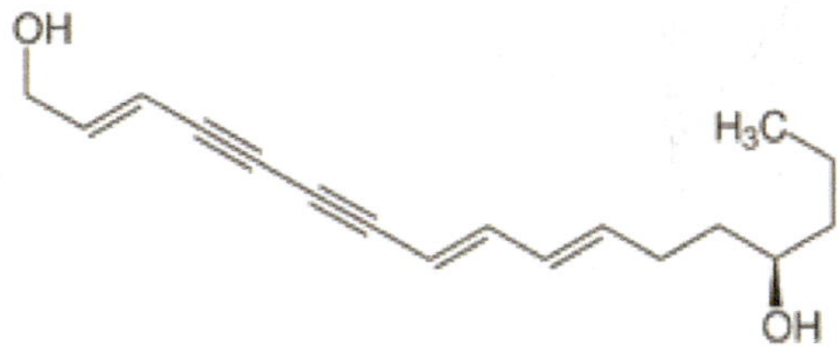

Oenanthotoxin

Apocynum cannabinum

1

2

Image credits : 1. Thayne Tuason, ,Wikipedia ; 2.Ryan Hodnett ,Wikipedia

Family: Apocynaceae

Common name (s) : Dogbanea

Global distribution: It is found distributed throughout the United States

Ecology : It occurs along ditches and waterways

Description: It is a perennial forb (flowering herb) with simple and opposite leaves which are with entire margins and without stipules. Flowers are mall flowers with white, pink, or greenish petals and are borne in terminal, compound cymes;

Harmful effects: The toxicity of this plant is due to its cardiotonic glycosides viz. cimarin and strophanthidin which also possess emetic and diuretic activity. The signs of poisoning include diarrhea with blood, cardiac insufficiency, bradycardia, and arrhythmias.

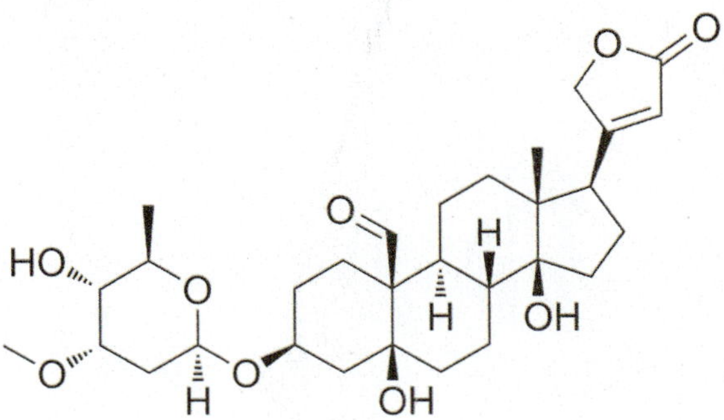

Cymarin

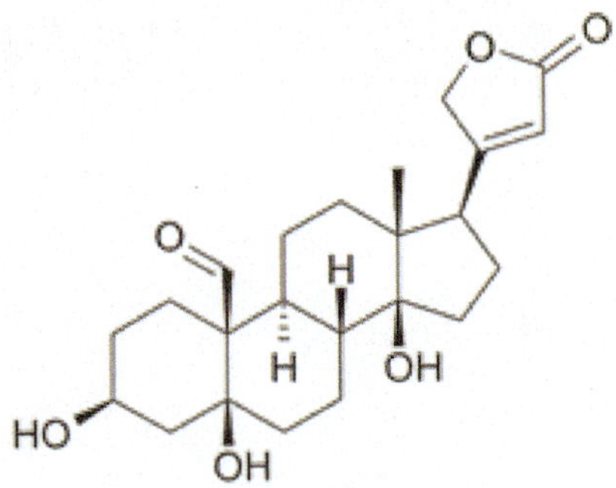

Strophanthidin

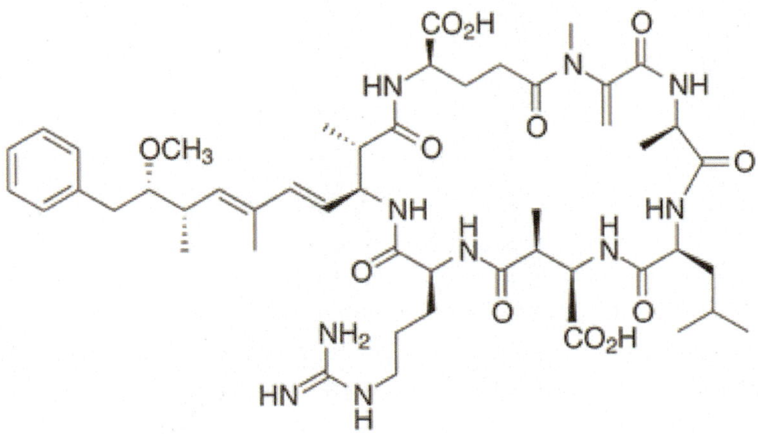

Cyanotoxin

Nerium oleander

Image credit: Wikipedia

Family: Apocynaceae

Common name: Oleander or Nerium oleander

Global distribution: It is native to Europe and Asia.

Ecology: It is common in river banks in moist soil

Description/ Biology: It is a shrub or small tree. Its stem is green, stout and shiny. Leaves are opposite or in whorls of 3 or 4. Its flowers are 5-petaled, and are in cymes. Fruit is a pair of follicles which have splits on the seam.

Harmful effects: It is the most poisonous plant in the world. All parts of the plant contain two types of most potent toxins viz. oleandrin and nerine which are known to be highly toxic and they may even be fatal to humans by affecting their hearts (https://adventure.howstuffworks.com/top-5-poisonous-plants10.htm; https://plants.ces.ncsu.edu/plants/nerium-oleander/)

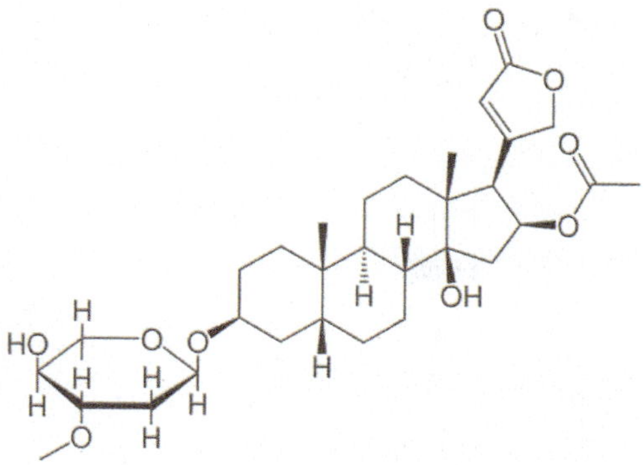

Oleandrin

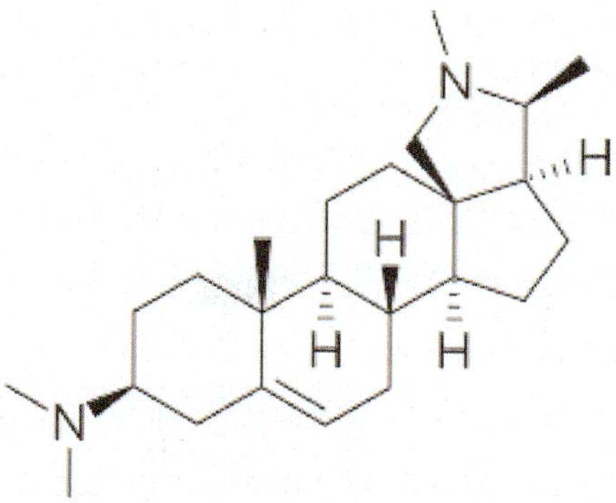

Nerine

Peltandra virginica

Image credit: Miguel Andrade , Wikipedia

Family: Araceae

Common name: Arrow arum

Global distribution: Eastern United States and Quebec, Ontario, and Cuba; central Florida including Everglades and Gulf Coast.

Ecology: It occurs in wetlands and swamps, including shallow waters , marshes and bogs

Description/ Biology: It has large, triangular leaves which appear above water. Leaves have protrusions near the stems which make them look like arrows, hence their name. These plants have fleshy, cylindrical inflorescence.

Harmful effects: The roots and foliage of this plant contain crystals of calcium oxalate which cause pain in the mouth, throat, or stomach if consumed (https://www.gardenguides.com/127812-poisonous-water-plants.html).

Pistia stratiotes

Image credit: Wikidata

Family: Araceae

Common name: Water lettuce

Global distribution: Water lettuce is considered an invasive pest species into Texas, Arizona and California via the southern United States.

Ecology: It is a floating aquatica plant.

Description/ Biology: It has pale green leaves, which grow in a spiral pattern from the center of the plant. These leaves grow to about 15 cm in length and are with deep ridges or depressions. It is an aggressive surface grower,

Harmful effects: The leaves of this plant contain poisonous calcium oxalate crystals, which may cause illness if ingested or chewed, especially by small animals and children. Symptoms include swelling and severe burning of the throat, lips and tongue, vomiting, nausea and diarrhea. In severe cases, the throat may swell leading to respiratory failure (https://www.ehow.com/list_6628387_poisonous-water-plants.htm; https://plants.ces.ncsu.edu/plants/pistia-stratiotes/). This plant by its fast growth may also kill fish by depleting the oxygen content the water bodies

Asclepias verticillata and *A. subverticillata*

A. verticillata (1)

A. subverticillata (2)

Image credit: 1. Mason Brock (Masebrock) , Commons Wikimedia; 2. Braderfly, Commons Wikimedia

Family: Asclepiadaceae

Common name (s) : Whorled milkweed and horsetail milkweed respectively

Global distribution: USA

Ecology : It is commonly seen in stream banks

Description: Both the species are similar in appearance and reach a height of 120 cm. leaves are linear shaped and narrow and are found as a whorl around the stem. Flowers are cream-colored and are found in groups which possess umbrella-like heads at the top of the stem.

Harmful effects: The types of poisoning associated with these species are neurologic type. and digestive tract/cardiac type. The pregnane glycosides viz cynafosides of this plant have been reported to be responsible for neurologic type poisoning. Unintentional consumption of this plant may yield symptoms such as trembling, ataxia, breathing difficulty, dilated pupils and coma followed by death due to respiratory failure. While the symptoms appear within a few hrs of consumption, death occurs in a day or two (Allison, https://pubs.nmsu.edu/_circulars/CR678/)

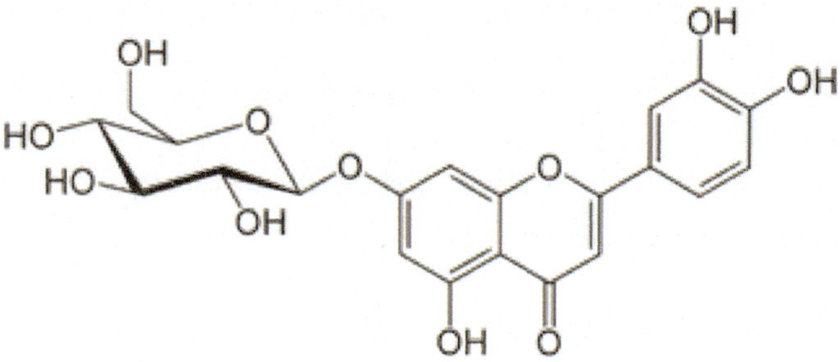

Cynafoside

Convallaria majalis

Image credit: H. Zell , Wikipedia

Family: Asparagaceae

Common name: Lily of the valley

Global distribution: It is native throughout the cool temperate Northern Hemisphere in Asia and Europe.

Ecology: it occurs mainly in ponds

Description/ Biology: The stems of this plant grow to a maximum height of 30 cm with one or two leaves of 10–25 cm long. Flowering stems however, have two leaves and araceme of 5-15 aflowersaon the stem apex. Flowers have six whiteatepalsa(rarely pink) which are fused at the base to form a bell-shape.

Harmful effects : This plant is toxic due to its glycosides. Ingestion of any part of the plant can cause serious illness or death, particularly in children. Ingestion of large amounts of the plant may result in abdominal pain, vomiting, diarrhea, slow or irregular pulse, burning of the mouth and throat, dizziness, dilated pupils, clammy skin, coma, circulatory or heart failure. (https://www.ehow.com/list_6628387_poisonous-water-plants.html; https://en.wikipedia.org/wiki/Lily_of_the_valley)

Helenium autumnale

Image credit: Commons Wikimedia

Family: Asteraceae.

Common name (s) : Common sneezeweed and flowered sneezeweed

Global distribution: North America

Ecology : It is commonly seen in ditches and bogs.

Description: It is a perennial herb growing to maximum height of 200 cm. Leaves are simple, alternate, elliptic to lanceolate and serrate to entire. Flowers are yellow and are seen in distinct heads.

Harmful effects: The toxicity of this plant is due to a glycoside, dugaldin, and a phenol. Livestock such as cattle and sheep have been reported to be largely affected by this plant. The symptoms associated with this plant include vomiting and inhalation pneumonia. Animals invariably die due to mainly secondary ailments.

Isocoma tenuisecta

Image credit: Wikipedia

Family: Asteraceae

Common name (s) : Shrine jimmyweed, and burrow goldenweed

Global distribution: It is native to Arizona, New Mexico, and Sonora.

Ecology : It inhabits largely in river drainages and irrigation canals.

Description: This half-shrub may grow to a maximum height of about 120 cm. Leaves which are are resinous and sticky are narrow and alternate and are divided into linear lobes . Its yellow flowers are borne in flat-topped clusters;

Harmful effects: The toxicity of this species is due to its compound tremetone, or tremetol which is largely seen in the green and dry leaves. The cumulative poison of this plant has been reported to be transmitted to humans through the milk of the cow, causing milk sickness. In humans, its toxin impairs the tricarboxylic acid (TCA) cycle and glucose utilization.The symptoms of the affected animals include muscular trembling, vomiting, labored breathing, constipation and dribbling of urine.

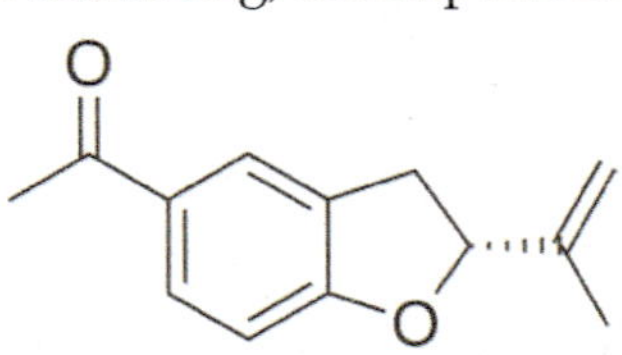

Tremetone

Oxytenia acerosa

Family: Asteraceae

Common name (s) : Copperweed

Global distribution: It is distributed in the western and northwestern areas of the USA

Ecology : It is also seen in valleys.

Description: This perennial and slender-stemmed erect plant has a maximum height of about 180 cm. Leaves are narrow and threadlike. Flowers are seen in dense panicles with many white discoid heads.

Harmful effects: The toxicity of this plant is often associated with the presence of a sesquiterpene alkaloid viz. pseudoguaianolide. Cattle are often affected by the poisoning of this plant. The symptoms of its toxicity include nervousness and anorexia followed by coma (Allison, https://pubs.nmsu.edu/_circulars/CR678/)

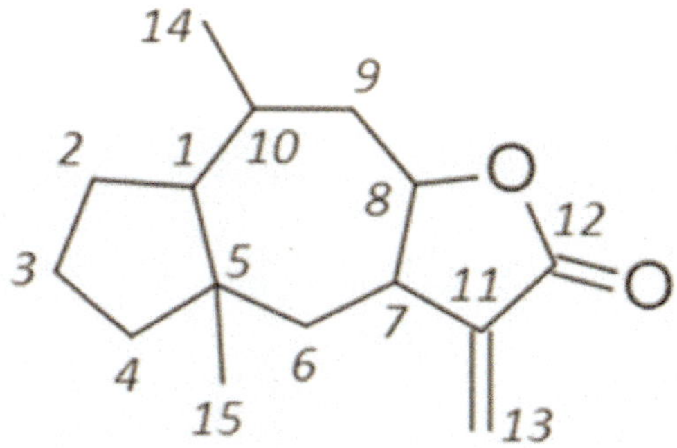

Pseudoguaianolide

Calycanthus floridus

Image credit: Yera , Wikipedia

Family: Calycanthaceae.

Common name (s) : Eastern sweet shrub and spice bush,

Harmful effects: The toxicity of these poisonous plants is largely due to their Lupinine and other quinolizidine alkaloids. Unintentional consumption of these plants has been reported to cause difficulties in breathing, trembling, convulsions, and coma followed by death. (Allison, https://pubs.nmsu.edu/_circulars/CR678/)

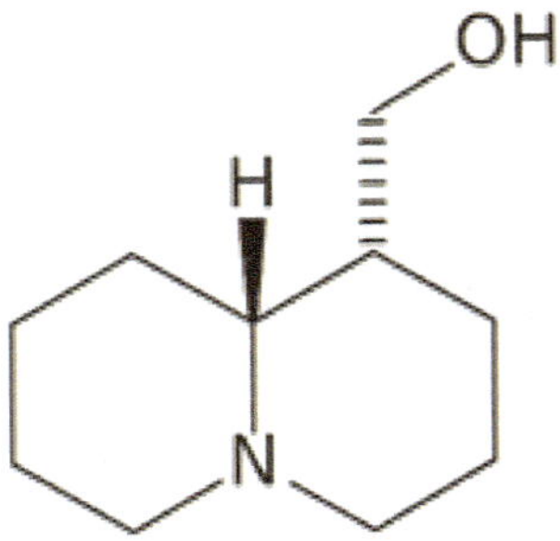

Lupinine

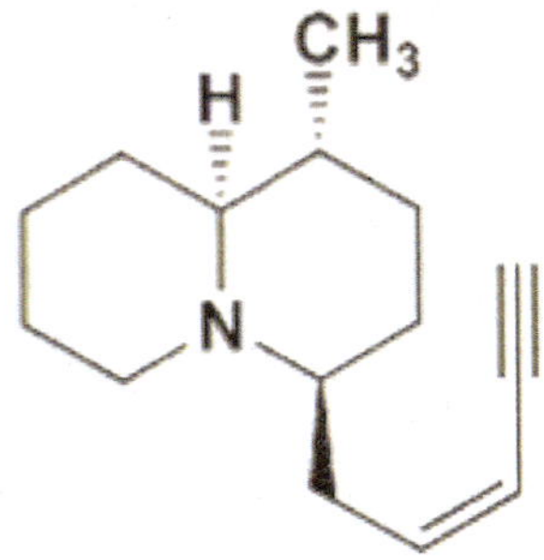

Quinolizidine alkaloid

Sesbania exaltata

Family: Fabaceae

Common name (s) : Colorado river hemp

Global distribution: North America

Ecology : It is also seen on stream banks

Description: It is an annual herb with a maximum height of 200 cm. Leaves are alternate and entire; and leaflets numbering 20-70 are oblong to linear elliptic. Flowers are yellow and are seen in axillary clusters of 2 to 6 flowers each.

Harmful effects: This poisonous plant which contains the glucoside viz. saponin has been reported to be toxic to both livestock and humans. The affected livestock develop symptoms which include hemorrhagic inflammation of the abomasum and intestines. They become comatose before death (https://carteret.ces.ncsu.edu/wp-content/uploads/2013/05/Poisonour-Plants-of-the-Southern-United-States.pdf?fwd=no)

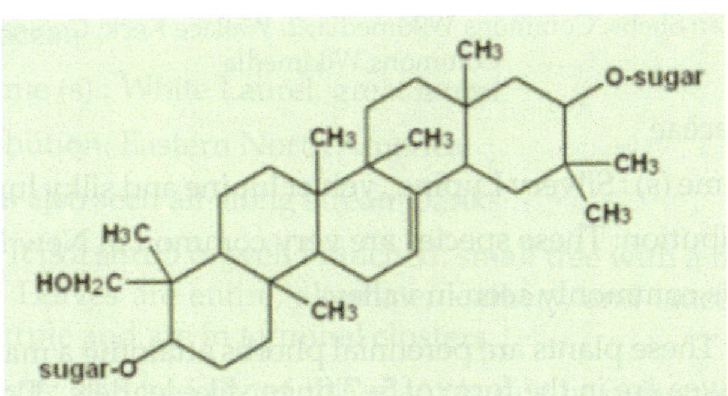

Saponin

Ecology : It is also seen in valleys.

Description: This perennial and slender-stemmed erect plant has a maximum height of about 180 cm. Leaves are narrow and threadlike. Flowers are seen in dense panicles with many white discoid heads.

Harmful effects: The toxicity of this plant is often associated with the presence of a sesquiterpene alkaloid viz. pseudoguaianolide. Cattle are often affected by the poisoning of this plant. The symptoms of its toxicity include nervousness and anorexia followed by coma (Allison, https://pubs.nmsu.edu/_circulars/CR678/)

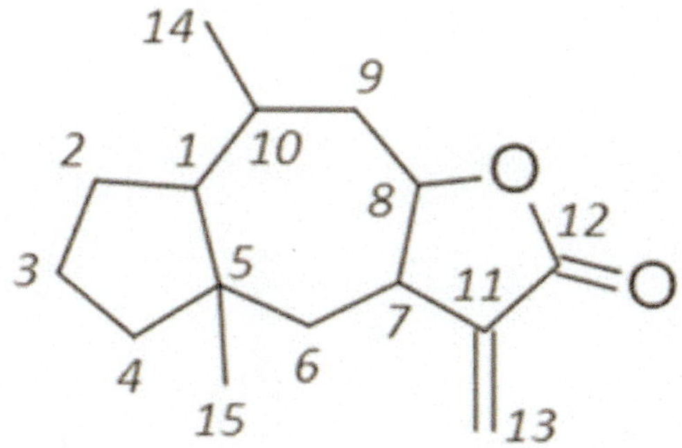

Pseudoguaianolide

Calycanthus floridus

Image credit: Yera , Wikipedia

Family: Calycanthaceae.

Common name (s) : Eastern sweet shrub and spice bush,

Global distribution: It is commonly distributed in the Southeastern United States

Ecology : It occurs along stream banks

Description: It is a shrub with opposite branches and is growing to a maximum height of only about 10 cm. Leaves are opposite, entire, and lanceolate to ovate. Its fragrant flowers are axillary with several linear petals and sepals.

Harmful effects: The toxicity of the seeds of this plant is due to an alkaloid which possesses pesticide characteristics. Affected cattle have been reported to exhibit tetanic convulsions before death (https://carteret.ces.ncsu.edu/wp-content/uploads/2013/05/Poisonour-Plants-of-the-Southern-United-States.pdf?fwd=no)

Lobelia cardinalis

Image credit: Barnes, Dr. Thomas G, Wikipedia

Family: Campanulaceae

Common name: Cardinal flower

Global distribution: Northern America

Ecology: It occurs in freshwater marshes, and all along the banks of rivers, streams and ponds.

Description/ Biology: Cardinal flower plant is perennial which grows to a maximum height of 180 cm. Its leaves are finely toothed and lance-shaped. It has small, red, tubular flowers with two lips and three lobes. Flowers bloom in July to September on alternate sides along unbranched stalks.

Harmful effects: All parts of the plant including the root are potentially toxic to humans. Its toxicity emerges only after eating large quantities of the herb or root. Symptoms of poisoning include nausea, vomiting, diarrhea and lethargy (https://www.gardenguides.com/127812-poisonous-water-plants.html)

Halogeton glomeratus

Image credit: Sheri Hagwood , Wikipedia

Family: Chenopodiaceae

Common name (s) : Halogetona

Global distribution: Russia, China and USA

Ecology : It is commonly seen along rivers

Description: It is an annual herb with succulent, branching stems. Leaves are succulent and tubular with weak spine at the tips. Flowers are inconspicuous and greenish in coloration.

Harmful effects: The toxicity of this plant is due to its sodium and potassium oxalates. Symptoms associated with this unintentional consumption include rapid respiration, stiffness, ataxia, and coma (Allison, https://pubs.nmsu.edu/_circulars/CR678/)

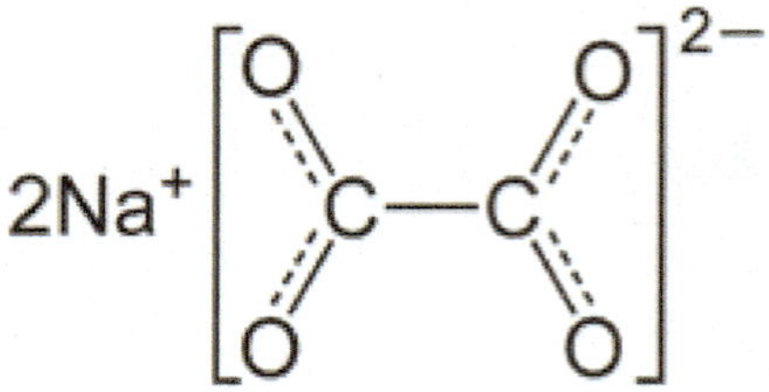

Sodium oxalate

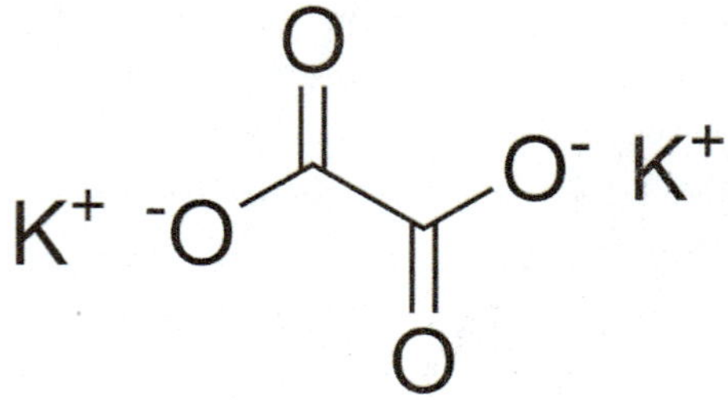

Potassium oxalate

Suckleya suckleyana

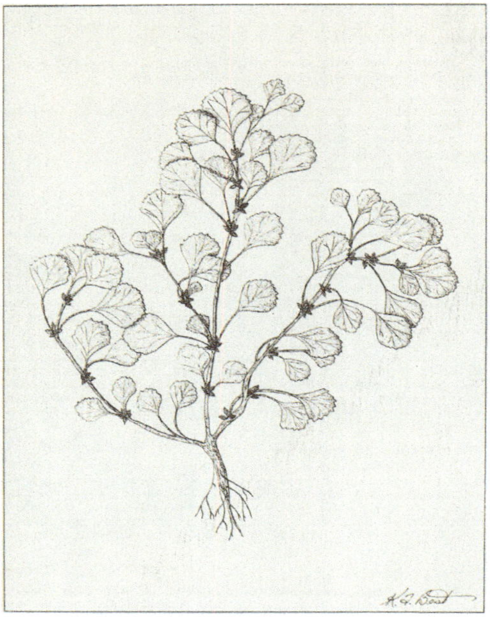

Image credit: Internet Archive Book Images, Wikipedia

Family: Chenopodiaceae

Common name (s) : Poison suckleya

Global distribution: Northeastern New Mexico

Ecology : It is seen usually around ponds

Description: It is a succulent plant with reddish stems of 30-60 cm long. Leaves are triangular shaped and alternate with dentate margins and long petioles. Flowers which are small and inconspicuous are borne in leaf axils. It is a monoecious plant with male flowers located on the tips of the branches and female flowers are present in other branching stems along the remainder of the branching stems;

Harmful effects: The toxicity of this poisonous plant is due to the presence of cyanogenic glycosides at a level of 100-400 ppm. Risk of poisoning is believed to increase if affected people drink water after consumption of the plant unknowingly. Acute poisoning is associated with symptoms such as distress, muscle twitching, dribbling of urine and staggering. In affected people, the respiratory rate increases initially and heart rate becomes irregular and rapid and death occurs a few minutes after these symptoms (Allison, https://pubs.nmsu.edu/_circulars/CR678/)

Lyonia ligustrina

Image credit: Eric Hunt , Wikipedia

Family: Ericaceae

Common name (s) : Maleberry, he-huckleberry

Global distribution: It occurs throughout the southern Unites States east of the Mississippi River

Ecology : It occurs in wetlandamarginsa

Description: This shrub grows to a maximum height of 300 cm. Leaves are alternate and obovate to elliptic. Flowers are globular and white and are borne in clusters at the end of branches.

Harmful effects: The toxicity of this plant is due to its resinoid, andromedotoxin, and a glucoside, arbutin. Cases of toxicity and deaths have been recorded in humans. Symptoms associated with the unintentional eating of the leaves of this plant include vomiting and muscular spasms and coma (https://carteret.ces.ncsu.edu/wp-content/uploads/2013/05/Poisonour-Plants-of-the-Southern-United-States.pdf?fwd=no)

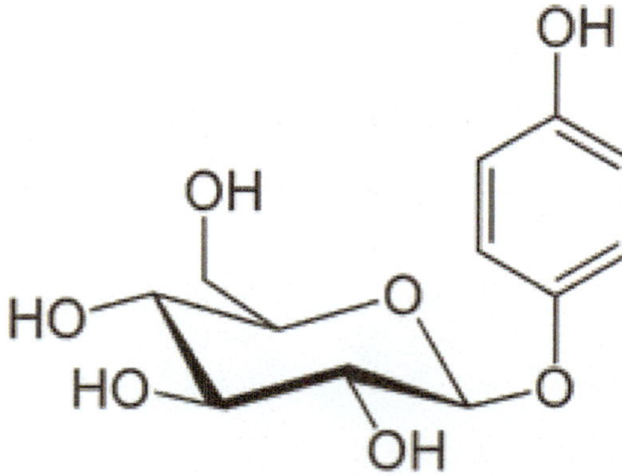

Arbutin

Rhododendron maxima

Image credit: Arx Fortis, Wikipedia

Family: Ericaceae

Common name (s) : White Laurel, great laurel

Global distribution: Eastern North America

Ecology : It is also seen all along stream banks

Description: It is a shrub or well branched, small tree with a maximum height of about 80 cm . Leaves are entire, alternate, leathery, and lanceolate to elliptic. Flowers are to purple and are in terminal clusters.

Harmful effects: The toxicity of this plant is due to poisonous carbohydrate viz. andromedotoxin or glucoside of hydroquinone. Deaths have been recorded due to the unintentional eating of this plants in general and leaves in particular. Symptoms associated with this toxic plant include vomiting, diarrhea, abdominal pain, paralysis of arms and legs , cardiac failure, difficult breathing, etc. (https://plants.ces.ncsu.edu/plants/rhododendron-maximum/; https://carteret.ces.ncsu.edu/wp-content/uploads/2013/05/Poisonour-Plants-of-the-Southern-United-States.pdf?fwd=no)

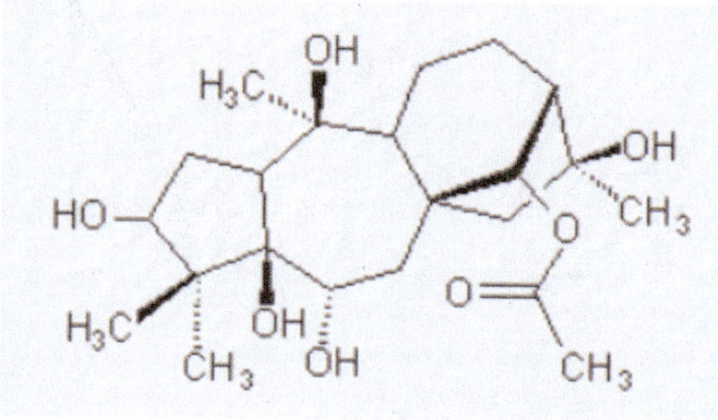

Andromedotoxin

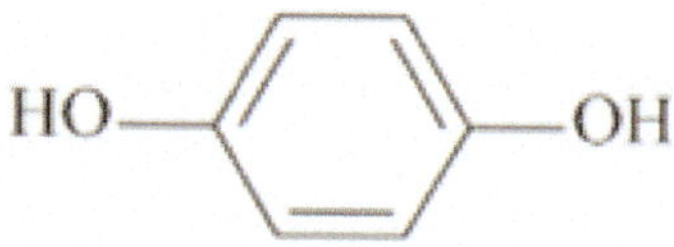

Hydroquinone

Lupinus argenteus, L. leucophyllus* and *L. sericeus

***L. argenteus* (1)**

***L. leucophyllus* (2)**

***L. sericeus* (3)**

Image credit: 1. Stan Shebs, Commons Wikimedia; 2. Wallace Keck, Commons Wikimedia; 3. Commons Wikimedia

Family: Fabaceae

Common name (s) : Silvery Lupine , velvet lupine and silky lupine respectively

Global distribution: These species are very common in New Mexico of USA

Ecology : It is commonly seen in valleys

Description: These plants are perennial phorbs attaining a maximum height of about 90 cm. Leaves are in the form of 5–7 finger-like leaflets . Flowers are seen in loose clusters surrounding the upper stalks and are generally blue colored

Harmful effects: The toxicity of these poisonous plants is largely due to their Lupinine and other quinolizidine alkaloids. Unintentional consumption of these plants has been reported to cause difficulties in breathing, trembling, convulsions, and coma followed by death. (Allison, https://pubs.nmsu.edu/_circulars/CR678/)

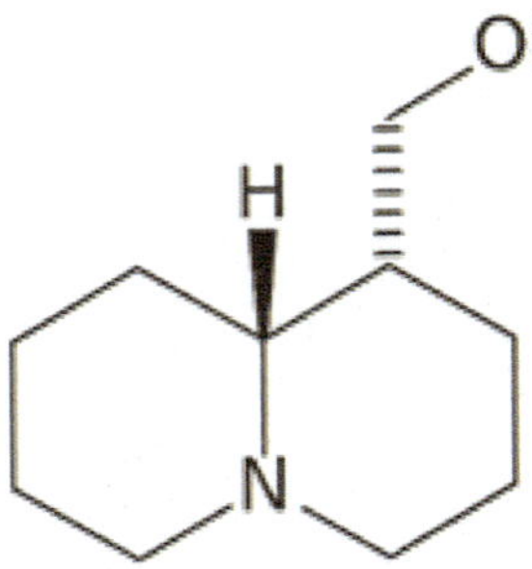

Lupinine

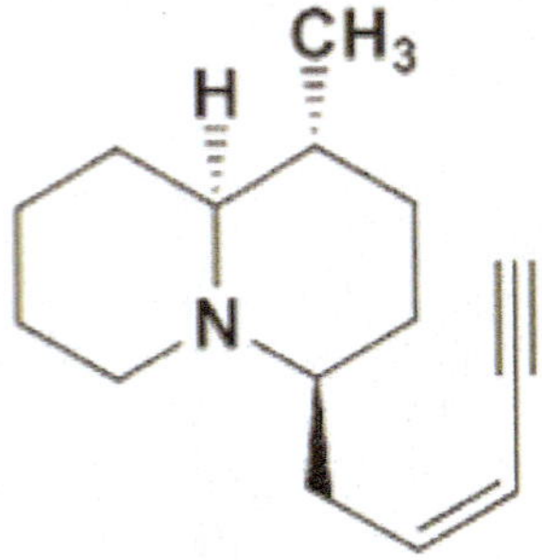

Quinolizidine alkaloid

Sesbania exaltata

Family: Fabaceae

Common name (s) : Colorado river hemp

Global distribution: North America

Ecology : It is also seen on stream banks

Description: It is an annual herb with a maximum height of 200 cm. Leaves are alternate and entire; and leaflets numbering 20-70 are oblong to linear elliptic. Flowers are yellow and are seen in axillary clusters of 2 to 6 flowers each.

Harmful effects: This poisonous plant which contains the glucoside viz. saponin has been reported to be toxic to both livestock and humans. The affected livestock develop symptoms which include hemorrhagic inflammation of the abomasum and intestines. They become comatose before death (https://carteret.ces.ncsu.edu/wp-content/uploads/2013/05/Poisonour-Plants-of-the-Southern-United-States.pdf?fwd=no)

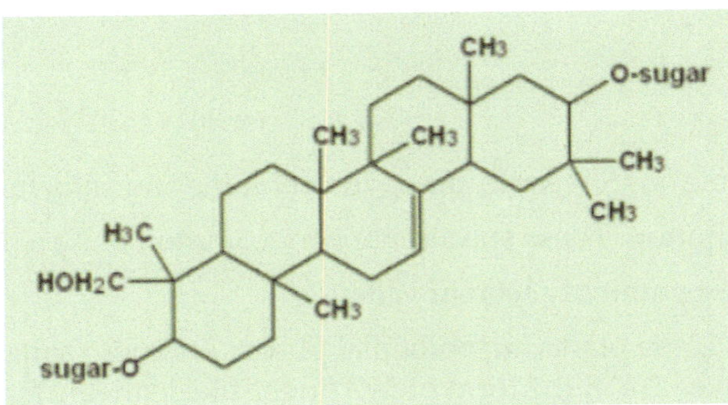

Saponin

Iris pseudacorus

Image credit: Photo: Myrabella , Wikipedia

Family: Iridaceae

Common name: Yellow flag, yellow iris or water flag

Global distribution: It is native to Europe, western Asia and northwest Africa

Ecology: It is an emergent aquatic plant

Description/ Biology: It is a perennial, emergent aquatic plant growing to a maximum height of 1.5 m. Its inflorescence units consist of 4–12 flowers per stem. Its bright yellow flowers are approximately 7–9 cm wide.

Harmful effects: This plant is poisonous to grazing animals like cattle and swine. The glycosides of the leaves and rhizomes of this plant cause abdominal pain, gastroenteritis, nausea, vomiting, diarrhea, spasms, staggering, and paralysis in sheep and cattle. Further, caution should be exercised if pulling out this plant as it causes skin irritations (https://wiki.bugwood.org/Iris_pseudacorus)

Triglochin maritima and *Triglochin palustris*

Triglochin maritima

Triglochin palustris

Image credit: Commons Wikimedia

Family: Juncaginaceae

Common name (s) : Common arrowgrass

Global distribution: Throughout New Mexico

Ecology : It is commonly seen in freshwater marshes and bogs.

Description: This perennial , unbranched and grass-like plant grows to a maximum height of about 100 cm. Leaves are basal, linear and unjointed. Flowers are small, greenish and bunched together on the upper part of the main flowering stem.

Harmful effects: The toxicity of this species is due to mainly a tyrosine-derived cyanogenic glucoside, triglochinin Usually livestock are said to be affected by its non-cumulative toxicity. Symptoms associated with its poisoning include rapid breathing, dilation of eyes, , convulsions, and coma followed by death. Plants grown in drought conditions are said to be 5–10 times more toxic than plants in submerged condition.

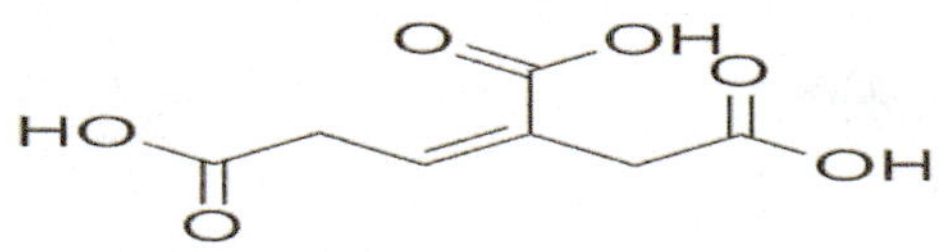

Triglochinin

Amianthum muscaetoxicum

Image credit: homeredwardprice - Fly Poison, Wikipedia

Family: Liliaceae

Common name (s) : Crow poison, fly Poison, stagger grass

Global distribution: It occurs throughout the southern Unites States east of the Mississippi River

Ecology : It is also seen in bogs

Description: This perennial, bulbous herb grows to a maximum height of about 150 cm. Leaves are linear and basal. Flowers are borne in dense racemes. They are normally white and are likely to change into greenish coloration in age as the plant grows.

Harmful effects: This is a very poisonous plant containing toxic alkaloids in all parts of the plant. (https://plants.ces.ncsu.edu/plants/amianthium-muscitoxicum/). Cattle and sheep are most commonly affected by this plant. The symptoms associated with the consumption of the plant include subnormal temperature and rapid respiration; and death results due to respiratory failure (https://carteret.ces.ncsu.edu/wp-content/uploads/2013/05/Poisonour-Plants-of-the-Southern-United-States.pdf?fwd=no)

Gelsemium sempervirens

Image credit: Kenpei, Wikipedia

Family: Loganiaceae

Common name (s) : Carolina jessamine, yellow jessamine, poor man's rope and evening trumpet flower

Global distribution: USA and Guatemala

Ecology : It is also seen in swampy areas

Description: It is an evergreen and a twining vine with a maximum length of about 7 m. Leaves are opposite and lanceolate with entire margins and acute to acuminate tips. Flowers are yellow, trumpet-shaped and fragrant.

Harmful effects: The toxicity of this plant is due to its alkaloids gelsemine, gelseminine and gelsemoidine. Humans have been reported to be poisoned from sucking the nectar from the flowers of this plant or from consuming honey derived from the flowers. Animals affected by this plant develop symptoms including. dilated eyes and convulsive movements. Death usually occurs soon after the ani Death may occur in such animals when they attain comatose condition (https://carteret.ces.ncsu.edu/wp-content/uploads/2013/05/Poisonour-Plants-of-the-Southern-United-States.pdf?fwd=no)

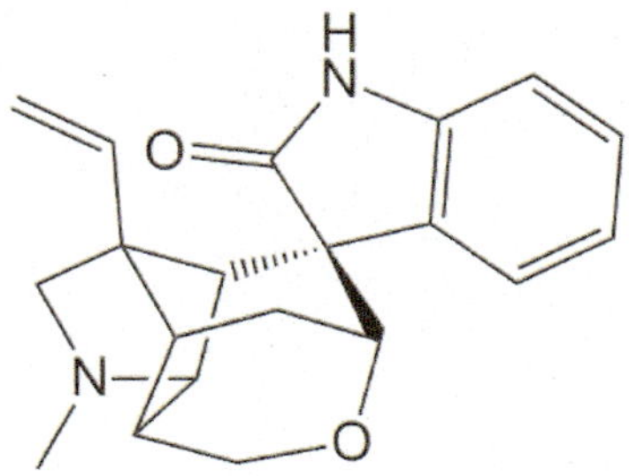

Gelsemine

Sorghum halepense

Image credit: Tau'olunga , Wikipedia

Family: Poaceae

Common name (s) : Johnson Grass

Global distribution: USA

Ecology : It is commonly seen in stream bottoms.

Description: It is a perennial plant with a maximum height of 180 cm. Root stalks of this plant are scaly in appearance. Leaves are broad.

Harmful effects: The toxicity of this poisonous plant is due to its prussic acid (= hydrocyanic acid or cyanide) and very high nitrate content. The symptom associated with prussic acid poisoning is the appearance of bright red venous blood and that of nitrate poisoning is dark brown arterial and venous blood.

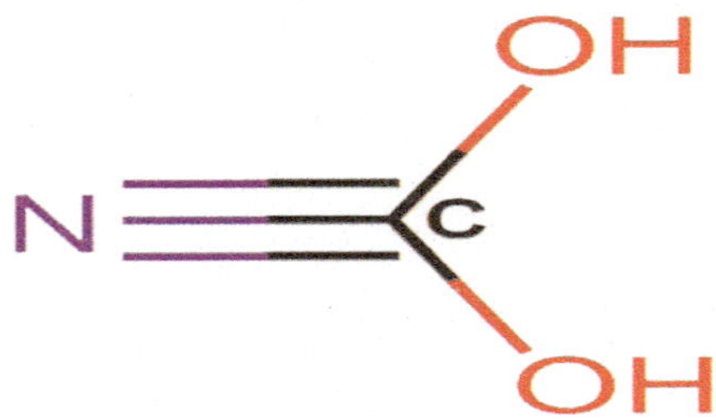

Hydrocyanic acid

Ledermanniella schlechteri

Family: Podostemaceae

Common name (s) : Not designated

Global distribution: Its native range is Ivory Coast of W. Central Tropical Africa

Ecology : It is a submerged freshwater herb growing on rocks of rapid waters. with ribbonlike roots which cover the substrate like a carpet.

Description: It is a perennial herb with long, branched, stems . Presence of inverted flowers is the characteristic feature of the species of this genus. .

Harmful effects: This plant is the most consumed aquatic vegetable in Africa. But the level of toxic metals in this species has been reported to exceed the FAO/ WHO permissible limits for human consumption. Health of humans eating this plant is likely to be impacted. The average values (in mg kg^{-1}) of toxic metals of this plant have been reported as ranging from 0.4 -9.1 (Cr), 0.1-4.5 (Ni), 5.5-78.4 (Cu), 336.1-1520.9 (Zn), 0.1-0.5 (As), 0.2-0.8 (Cd), 0.4-11.8 (Pb) and 0.02-0.2 (Hg) (Mata et al., 2020)

Delphinium bicolor, D. andersonii, D. barbeyi, and *D. glaucum*

D. bicolor (1)

D. andersonii (2)

D. barbeyi (3)

D.glaucum (4)

Image credit: 1. Matt Lavins, Commons Wikimedia; 2. Cassondra Skinner, commons. wikimedia; 3. Dave Powell, USDA Forest Service, commons.wikimedia; 4. Wikimedia

Family: Ranunculaceae

Common name (s) : Larkspurs

Global distribution: West and Northwestern United States

Ecology : Tall species are commonly seen in streams and valleys

Description: *Delphinium bicolor* and *D. andersonii* are short, perennial forbs with hairy stems and a maximum height of only 60 cm. In these species, only a few leaves which start from the cluster of roots are rounded and are well divided into several linear segments. *Delphinium barbeyi* and *D. glaucum* on the other hand are tall plants with a maximum height of 210 cm. In these species, the leaves are divided into deep lobes and are broad. Flowers are whitish, blue or purple. Along the main flowering stem, there may a maximum of 15 flowers on short stalks

Harmful effects: The toxicity of these poisonous plants is largely due to their diterpenoid alkaloids. Short species seem to possess even 27 alkaloids which are in the form of 19-C and 20-C terpenes and bis-diterpenes. In the tall species, the 19-C diterpenes are the aconitine and lycoctonine types. Symptoms associated with the poisoning include stiffness , nervousness and staggering; and death invariably results due to respiratory and heart failure.

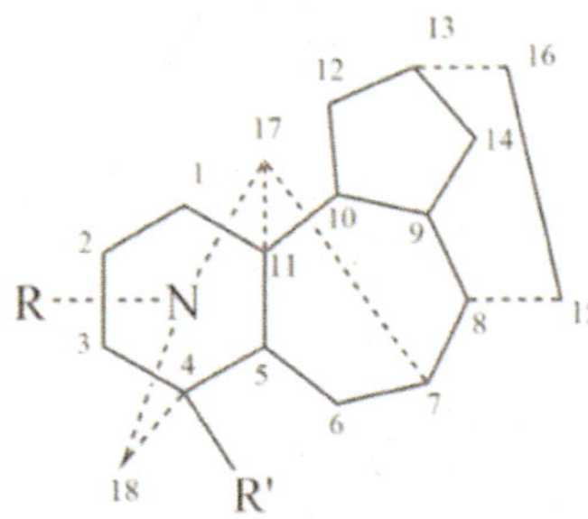

Diterpenoid alkaloid

Ranunculus abortivus

Image credit: Jay Stumer, Wikimedia commons

Family: Ranunculaceae

Common name (s) :Little-leaf buttercup and kidney-leaf Buttercup

Global distribution: Eastern North America

Ecology : It is most common in the banks of rivers and ditches; and pond margins.

Description: It is an annual or perennial herb with alternate stem leaves which are simple, lobed or divided. . Flowers are solitary and axillary with five glossy, yellow petals.

Harmful effects: The stem and leaves of this plant contain a highly toxic, irritant oil called protoanemonin. Affected livestock have been reported to develop severe abdominal pain, diarrhea, and convulsions and death may result subsequently.)

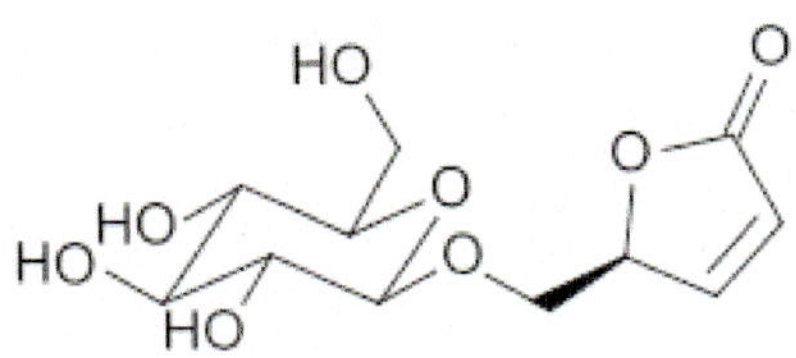

Protoanemonin

Prunus virginiana

Image credit: Nadiatalent , Wikipedia

Family: Rosaceae

Common name (s) : Chokecherry

Global distribution: It is found distributed throughout USA

Ecology : It inhabits riverbanks and valleys

Description: It is an erect shrub or tree growing up to a total length of about 600 cm. Leaves are oval and dark green. They have a pointed tip, rounded base, and saw-toothed margins. Flowers are small and yellow-white and are seen in dense clusters like a cylindrical raceme;

Harmful effects: The toxicity of this plant is due to the presence of two cyanogenic glycosides viz. amygdalin (in the seed)and prunasin (in the leaves and stems). Cattle are believed to be affected by this plant. The signs of poisoning of this plant in cattle include abnormal breathing , dilation of eyes, convulsions, and coma followed by death. Reports are also available on the possible poisoning of its fruits in humans (Allison, https://pubs.nmsu.edu/_circulars/CR678/)

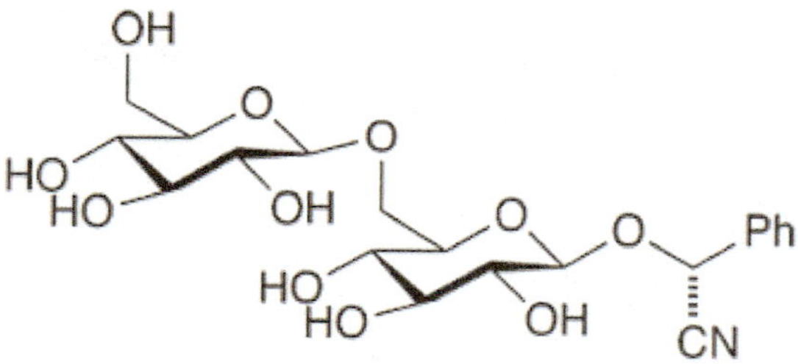

Amygdalin

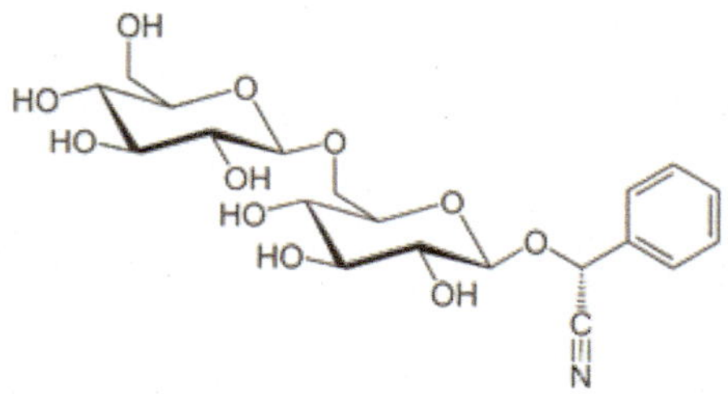

Prunasin

Cephalanthus occidentalis

Image credit: The Cosmonaut , Wikipedia

Family: Rubiaceae

Common name (s) : Buttonbush, button-willow, buck brush, and honey-bells

Global distribution: This species is native to the southern and eastern North America

Ecology : It occurs in the margins of lakes, rivers, creeks and swamps

Description: This shrub grows to a maximum height of 300 cm. Leaves are whorled or opposite and ovate to elliptic. Flowers are white and are borne in terminal or axillary spherical heads.

Harmful effects: The toxicity of this plant is largely due to the glycosides, cephalin and cephalothin. Cattles have been reported to be affected by the poisonous leaves of this plant. The symptoms associated with the eating of this plant are vomiting and convulsions ; and death invariably results.

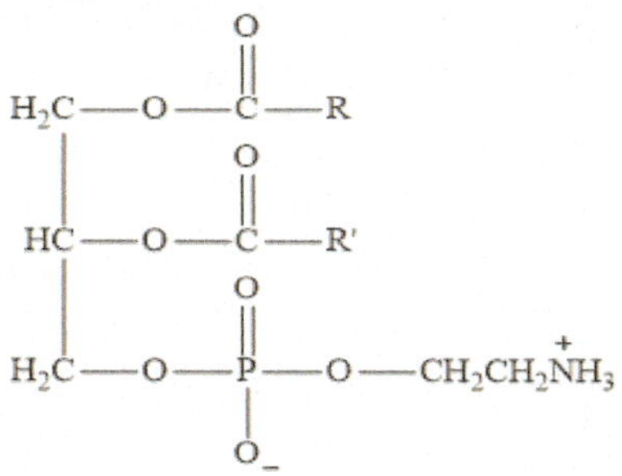

Cephalin

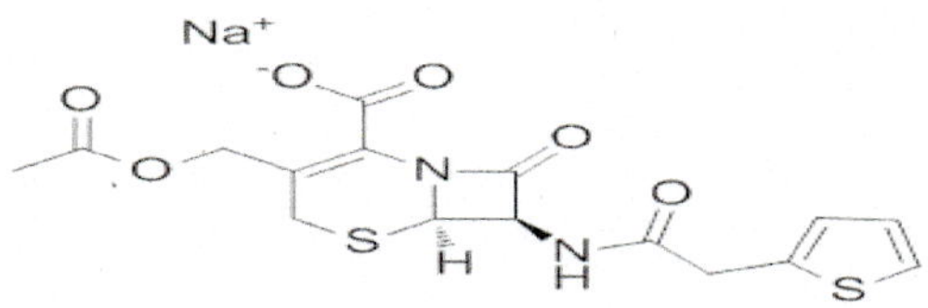

Cephalothin

Solanum dulcamara

Image credit: H.Zell, Wikipedia

Family: Solanaceae

Common name: Bitter nightshade, climbing nightshade

Global distribution: Throughout King County of Washington, USA

Ecology: It occurs in marshes, wetlands, lake shores, canal banks , forested freshwater dune barriers, and pond edges

Description/ Biology: Leaves of this plant are broadly ovate to ovate-lanceolate. Their lamina are 5.5–8 cm long and 3–4 cm wide and concolorous. Entire or upper leaves are lobed at base. Inflorescence is branched and is up to 20–flowered. Corolla is stellate and bluish with 10–12 mm diam.

Harmful effects: This plant is toxic due to its steroidal glycoalkaloid solanine . This plant may therefore be fatal if eaten. Following ingestion, solanine is poorly absorbed from the gastrointestinal tract causing clinical signs such as salivation, drowsiness , nausea, vomiting, abdominal pain, diarrhea, low blood pressure , weakness, respiratory depression, etc. (https://plants.ces.ncsu.edu/plants/solanum-dulcamara/#poison)

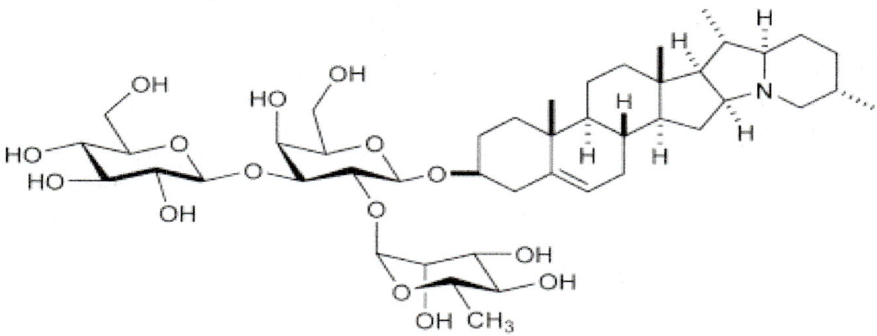

Solanine

CHAPTER 4

HARMFUL FRESHWATER INVERTEBRATES

4.1. Protozoans

Acanthamoeba spp.

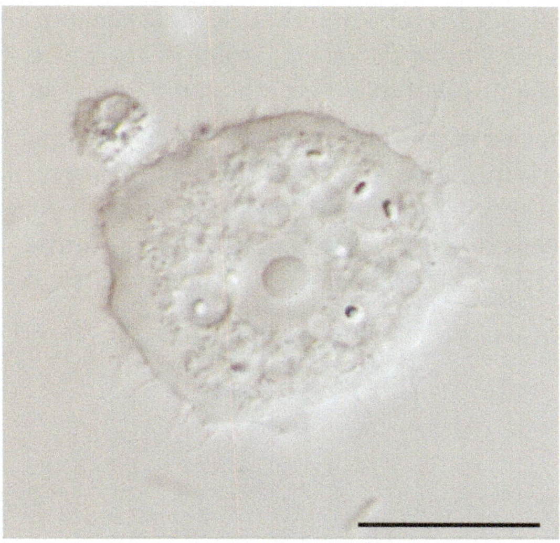

Image credit: Wikipedia

Phylum: Protozoa

Family: Acanthamoebidae

Global distribution: Worldwide

Habitat: The species of *Acanthamoeba* viz. *Acanthamoeba castellanii* , *Acanthamoeba polyphaga* and *Acanthamoeba culbertsoni* are free-living opportunistic protozoan pathogens and these organisms inhabit a variety of environments such as soil, and waters. These are amphizoic organisms i.e. they have the ability to exist both as free-living amebae and as parasitic pathogens

Description: The most characteristic feature of *Acanthamoeba* is the presence of spinelike pseudopods called acanthapodia. The different species of *Acanthamoeba* spp. have only the active feeding trophozoite and dormant cyst stages in their life cycle. Both the trophozoites and cysts may retain viable bacteria thereby serving as reservoirs for bacteria with human pathogenic potential.

Harmful organs and Harm: The aforesaid species of *Acanthamoeba* have been reported to be the causative agents of granulomatous amebic encephalitis (GAE), a fatal disease of the central nervous system affecting the brain and spinal cord; and amebic keratitis (AK) and have normally been associated with cutaneous lesions and sinusitis. GAE is a serious infection of the brain and spinal cord which occurs in AIDS patients and other immunocompromised individuals. AK is a serious eye infection mainly affecting contact lens users and it occurs even in healthy persons resulting blindness (Marciano-Cabral and Cabral, 2003).

Ashbolt (2015) reported that the species of *Acanthamoeba* may also cause "meningoencephalitis" (an inflammation of the brain), a severe and life-threatening infection affecting immuno-competent children and immuno-compromised adults

Al-Herrawy et al. (2015) reported that *Acanthamoeba castellanii, Acanthamoeba culbertsoni,* and *Acanthamoeba polyphaga* isolated from the "finished" water of drinking water treatment plant may be pathogenic to humans.

The different species of this genus behave as environmental hosts of amoeba-resisting bacterial pathogens such as NTM (Nontuberculous mycobacteria) and legionellae. The species *Legionella pneumophila* has been reported to cause a pneumonia-type illness called "Legionnaires' disease" and a flu-like illness called "Pontiac fever" (Rodriguez-Zaragoza,1994; Ashbolt , 2015; Wannasan et al., 2013).

Balamuthia mandrillaris

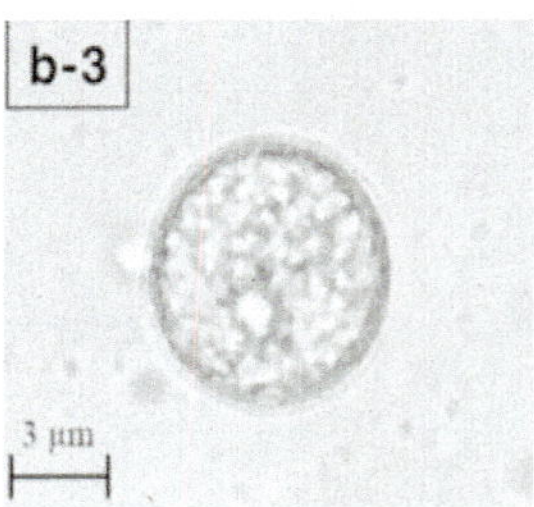

Image credit: Wikipedia

Protozoa Acanthamoebidae

Global distribution: It is distributed throughout the temperate regions of the world.

Habitat: It is a water-based (opportunistic) free-living amoeba (Ashbolt,2015). It may also be found in soil.

Description: It is a free-living, heterotrophic amoeba, consisting of organelles surrounded by a three-layered cell wall and with a large nucleus. Size of a trophozoite of this species is 30 - 120 µm in diameter. Cysts of this species however measure only 10-25 µm in diameter. These cysts have two walls: a wrinkled fibrous outer wall (exocyst) and an inner wall (endocyst) which may be hexagonal, spherical or polygonal. Cysts have only one nucleus with a large karyosome.

Harmful organs and Harm: This species can infect the human body through open wounds or by inhalation. On introduction, this amoeba may form a skin lesion, or may migrate to the brain. Cysts of this species were found in the brain, eyes, skin, lungs and other organs.

This species has been reported to cause a deadly neurological condition of the brain called "Granulomatous Amebic Encephalitis" (GAE) or "Granulomatous Balamuthia encephalitis" (BAE) in both immunocompetent and immunocompromised patients including children (Morsy et al., 2016; https://www.cdc.gov/parasites/balamuthia/index.html). Balamuthia-induced GAE can cause focal paralysis, seizures, and brainstem symptoms such as facial paralysis, headache , neck stiffness , difficulty in swallowing, and double vision. Since the time of the isolation of this pathogen, about 200 cases had been reported worldwide . In the United States, alone, 94 cases were reported from 1976 to 2014.

Balczun and Scheid (2017) reported that 100 cases had been reported, and only three survivors were known. The infection due to BAE is chronic and the time between infection and appearance of neurological symptoms normally ranges from one month to about two year.

Protacanthamoeba sp.

Family: Acanthamoebidae

Habitat: It dwells both in freshwater and soil. It is an amphizoic and exists either as a parasite or as a free-living amoeba.

Description: Its locomotive form has a size range of 20 - 40 µm. Nucleus is with a central nucleolus. Cysts of this species are smooth and circular to oval, and has a size of 16.5 µm across.

Harmful organs and Harm: This species has been reported to behave as an environmental host of amoeba-resisting bacterial pathogens such as NTM (Nontuberculous mycobacteria) and legionellae. The species *Legionella pneumophila* has been reported to cause a pneumonia-type illness called "Legionnaires' disease" and a flu-like illness called "Pontiac fever".

Blastocystis hominis

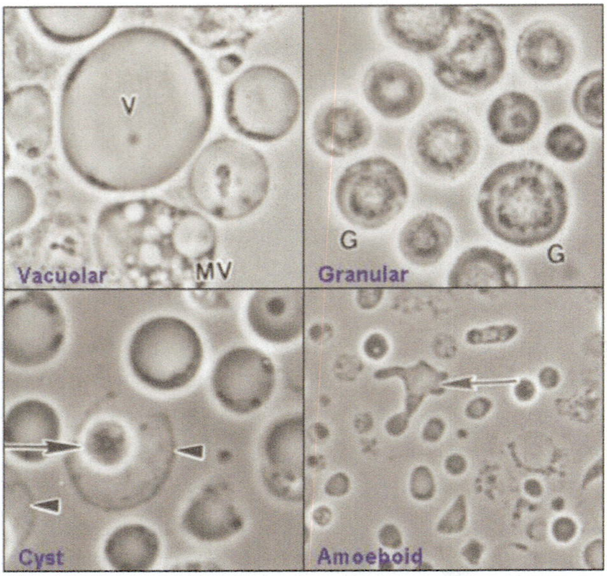

Common forms of Blastocystis hominis - vacuolar, granular, amoeboid, and cyst

Image credit: Wikipedia

Family: Blastocystidae

Global distribution: It is found worldwide.

Habitat: It is an enteric (waterborne) pathogen (Ashbolt,2015) and is normally found in the stools of people who have ingested water from contaminated freshwater bodies.

Description: This microscopic organism is found mainly as a vacuolar (or central body) form and as an infectious cyst form. While the size of the former stage varies from 5 to 40 µm, cyst stage is between 3 and 5 µm only. Other morphologic forms such as ameboid and granular forms have also been noted. Reproduction is by asexual method and is usually by binary fission. This organism has no cell wall and its major organelles include multiple nuclei, Golgi apparatus, mitochondria, endoplasmic reticulum, fat and other inclusions.

Harmful organs and Harm: This parasite was found to be the most common parasite in Indonesian patients with human immunodeficiency virus infection and AIDS who had chronic diarrhea. This pathogen causes gastrointestinal problems in the affected people and the symptoms associated with the infection include: watery diarrhea, nausea, abdominal pain, bloating, flatulence (excessive gas) , fatigue and loss of appetite

Cryptosporidium hominis

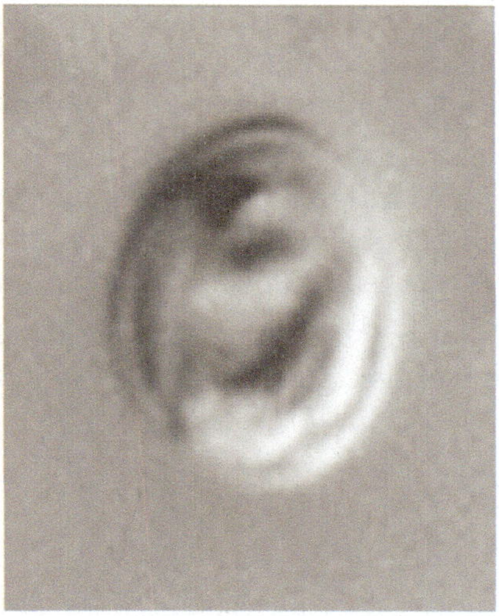

***Cryptosporidium* sp.**
Image credit: Centros para el Control y la Prevención de Enfermedades , Wikipedia
Phylum: Protozoa Family: Cryptosporidiidae

Global distribution: India, Peru, New Orleans, Malawi, South Africa, Kuwait, and Portugal

Habitat: It is an enteric (waterborne) species (Ashbolt, 2015) and it is spread through the fecal-oral route usually by drinking water contaminated with oocyst containing feces of humans (Xiao et al., 2004). Host of this species is only human.

Description: This parasite forms three developmental stages viz. meront, gamont and oocyst. Size of its oocyst is 4.9 x 4.3µm. The oocyst stage of this species has been reported to remain infective under cool, moist conditions for many months.

Harmful organs and Harm: Infection of this pathogen has only been reported from immuno-compromised patients, especially AIDS patients . This obligate parasite is responsible for the disease "cryptosporidiosis" in which the pathogen can colonize the gastrointestinal tract resulting in the gastroenteritis and diarrhea . Acute-chronic diarrhea may lead to considerable fluid and weight loss and even death Apart from the upper intestine, this parasite can also infect the stomach, pancreas, liver, and bile ducts of the patients. This pathogen has been reported to be the large contributors of child death and illness in heavily affected areas (Xiao et al., 2004; Hashim et al., 2006). Several reports of outbreaks of cryptosporidiosis have been related to drinking water contamination in North America, UK, and Japan (Fayer,2004).

Global distribution: Eastern states in the United States; China

Habitat: It is an enteric (waterborne) species (Ashbolt,2015) and It is spread through the fecal-oral route usually by drinking water contaminated with oocyst containing feces of humans (Xiao et al., 2004). Host of this species include mammals viz. humans, cattle, sheep, goats, horses, pigs and mice (Hashim et al., 2006).

Description: This parasite forms three developmental stages viz. meront, gamont and oocyst. Size of its oocyst is 5.0 x 4.5µm. The oocyst stage of this species has been reported to remain infective under cool, moist conditions for many months.

Harmful organs and Harm: It is the most frequently reported zoonotic species. Its infection largely resembles its congener, *Cryptosporidium hominis*.

Echinamoeba exundans

Family: Echinamoebidae

Global distribution: It has worldwide distribution

Habitat: All aquatic sources, e.g. lake, rivers , hot water systems and hot springs

Description: The trophozoite of this free living amoeba is oval to flattened, or elongated in shape. It has thin to spiny pseudopodia. Cysts are rounded and double-walled with one nucleus and a central karyosome or small, rounded with unstained nucleus.

Harmful organs and Harm: This species behaves as an environmental host of amoeba-resisting bacterial pathogens such as NTM (Nontuberculous mycobacteria) and legionellae. The species *Legionella pneumophila* has been reported to cause a pneumonia-type illness called "Legionnaires' disease " and a flu-like illness called "Pontiac fever" (Rodriguez-Zaragoza,1994; Ashbolt , 2015; Wannasan et al., 2013). Waterbodies contaminated with this pathogenic protozoan may cause a number of silent morbidities and mortalities particularly among rural households.

Thomas et al. (2008) reported that the species of this genus isolated from ozone-treated water served as a host for the pathogenic bacterial species, *Mycobacterium mucogenicum* which has been implicated in catheter related bloodstream infections

Cyclospora cayetanensis

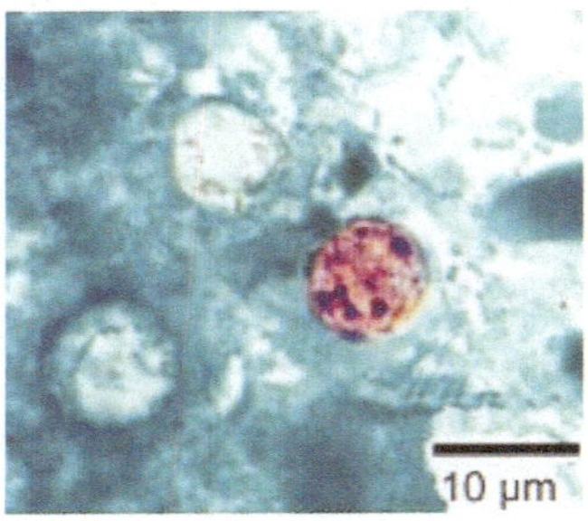

Image credit: Wikipedia

Family: Eimeriidae

Global distribution: Infection of this species has been reported worldwide, in both developed and developing countries.

Habitat: It is an enteric (waterborne) pathogen (Ashbolt,2015) and its only confirmed hosts are humans.

Description: It is a cyst-forming coccidian protozoan and its sporocysts are spherical in shape with a size range of 7.5 - 10 µm in diameter. The oocyst formula is O.2.2 as one oocyst contains two sporocysts and each sporocyst contains two sporozoites.

Harmful organs and Harm: It has been reported to cause a diarrheal disease called "cyclosporiasis" in humans and possibly in other primates. This protozoan lives out its lifecycle intracellularly within the human's epithelial cells and gastrointestinal tract. Infection is normally transmitted through the fecal-oral route, and begins when a person ingests oocysts in feces-contaminated water. The symptoms of its infection include watery diarrhea, loss of appetite, weight loss, flatulence, nausea, fatigue, and mild fever which may be enhanced in severe cases by muscle aches , vomiting, substantial weight loss, and chronic diarrhea (Ashbolt,2015).

Vermamoeba vermiformis **(=** *Hartmannella vermiformis*)

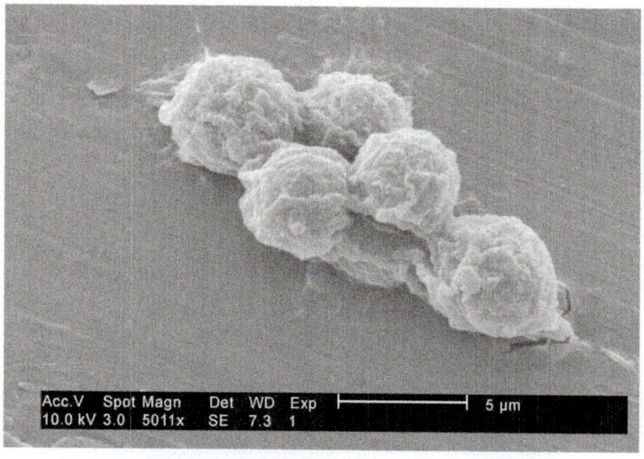

Image credit: Janice Haney Carr , Wikipedia

Family: Hartmannellidae

Global distribution: USA, UK and other European countries

Habitat: It is a water-based (opportunistic) (Ashbolt,2015). It occurs in natural freshwater environments and in thermal springs with temperatures of 41-53 °C. It acts as a vehicle and a reservoir of endocytobionts (i.e. bacteria, fungi, small protozoa or viruses) which are able to reside permanently or transiently in this species (Scheid, 2019).

Description: It is an ubiquitous, cyst-forming free-living amoeba showing resistance in different environmental conditions. It has a two-stage life cycle with a locomotive trophozoite and a dormant cyst. Trophozoites are worm-shaped or slug-like and are elongated and cylindrical. Locomotive forms are 22-42 µm long. Granulae are seen within the cytoplasm. In general, morphological features seem to vary among the strains, determined by the environmental conditions. Though the trophozoites are usually monopodial, they may exhibit several pseudopodia and irregular shapes, especially when they are moving in another direction. Cysts of this species are of spherical shaped with a diameter of 6-9µm. Its two-layered cyst wall contains proteins and glucose polymers; one or two nuclei and there are no pores or ostioles are visible (Scheid, 2019).

Harmful organs and Harm: The different species of this genus behave as environmental hosts of amoeba-resisting bacterial pathogens such as NTM (Nontuberculous mycobacteria) and legionellae. The species *Legionella pneumophila* has been reported to cause a pneumonia-type illness called "Legionnaires' disease " and a flu-like illness called "Pontiac fever" (Wikipedia; Rodriguez-Zaragoza,1994; Ashbolt , 2015; Wannasan et al., 2013).

A case of Acanthamoeba keratitis infection has also been reported due to this species .(Sente et al., 2016; Ashbolt , 2015).

Case reports: The medical significance and its pathogenic potential as a parasite for humans has been reported from several countries like Spain, Costa Rica, Peru, France, Scotland, Japan and Iran. In 2014, this species was detected in human nasal swabs . Most of the case reports relating to this species are however associated with corneal damage. Acanthamoeba keratitis case was reported in a female contact lens wearer, who initially had eye pain, redness, and blurred vision, Interestingly, this species has been reported predominantly in mixed amoebic keratitis cases (Scheid, 2019).

Giardia lamblia (= *Giardia intestinalis, Giardia duodenalis*)

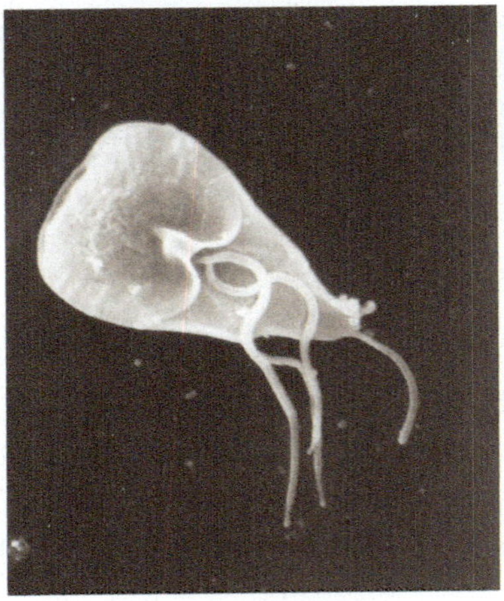

Image courtesy: Wikipedia

Family: Hexamitidae

Common name: Not reported

Global distribution: Throughout the world; and common in US, Soviet Union, Mexico, Southeast Asia, and western South America

Ecology: This enteric (waterborne) protozoan occurs commonly in stagnant water sources, such as naturally occurring ponds; and contaminated streams or lakes,

Description/ Biology: It is a flagellated unicellular eukaryotic microorganism with a pear shape. It has one or two transverse, claw-shaped median bodies. This species exists in two forms viz. an active form called a trophozoite, and an inactive form called a cyst.

Harm: The active trophozoite of this species attaches to the human small intestine and is responsible for causing the disease "giardiasis" which is popularly known as beaver fever The symptoms associated with this disease are diarrhea (watery diarrhea alternating with greasy stools), fatigue, cramps, belching gas, abdominal pain, and weight loss (Wikipedia). In certain areas of the world, water contaminated with G. lamblia cysts may causes travel-related giardiasis in tourists (Adam,2001).

Toxoplasma gondi

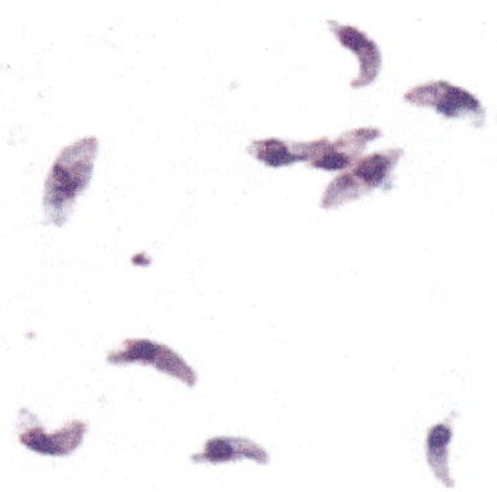

Image credit : Wikipedia

Family: Sarcocystidae

Global distribution: It is found worldwide and is one of the most common parasites in developed countries.

Habitat: It is an enteric (waterborne) pathogen (Ashbolt,2015). No studies are available on the ability of its lifecycle within freshwater environments, though infections are still present. It is believed that this species may be transmitted via other free-living species of amoeba such as , *Acanthamoeba* spp, which are found in all water bodies.

Description: This species contains organelles such as rhoptries and micronemes. Its lifecycle has two components viz. a sexual component that occurs only within cats (definitive host) and an asexual component which can occur in all warm-blooded animals, including humans (intermediate hosts).

Harmful organs and Harm: This species causes the infectious disease toxoplasmosis (fatal illness) . In humans, *T. gondii* is one of the most common parasites in developed countries. It has been reported that 30–50% of the global population has been chronically infected with this species. Highest prevalence of persons infected by this species was in France i.e. at 84% as on 2000. Although mild, flu-like symptoms may occur during the first few weeks of infection no readily observable symptoms were recorded in healthy human adults. However, severe infection may occur in infants, HIV/AIDS patients and others with weakened immunity (https://en.wikipedia.org/wiki/Toxoplasma_gondii).

Recent studies have suggested that the infection of this species may induce some of the alterations in the human brain as observed in mice. Further, infection may also be associated with behavioral changes and neurological disorders, such as schizophrenia, cognitive deficits and bipolar disorder. (https://en.wikipedia.org/wiki/Toxoplasma_gondii).

Sappinia diploidea

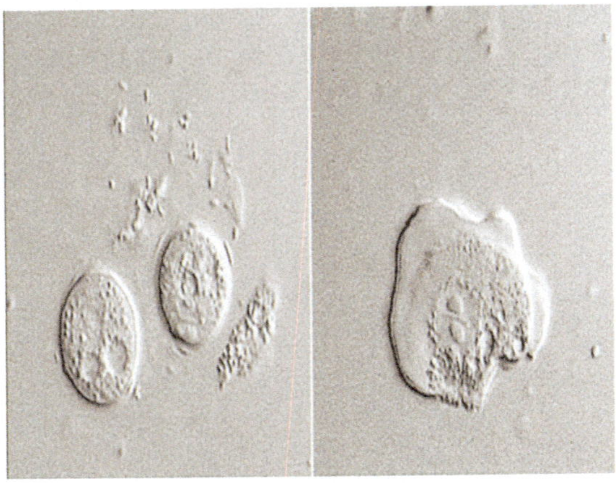

Image credit: Centers for Disease Control and Prevention (Reproduced with permission)

Family: Thecamoebidae

Global distribution: It has worldwide distribution

Habitat: It has been isolated from vertebrate feces, contaminated soil and fresh water

Description: Trophozoite (growing stage) of this free living amoeba is oval and binucleated with a size range of 40-70μM. The life cycle of *Sappinia* consists of two stages, a trophozoite and a cyst.

Harmful organs and Harm: This species has caused amoebic encephalitis which is an CNS infection of the brain (https://www.slideshare.net/DRABHISHEKJAIN2/free-living-amoebae). infected patient's brain also exhibited a 2-cm tumor-like mass on the back left section of his brain. Symptoms of this infection include headache, nausea, vomiting, blurry vision and loss of consciousness. The incubation period and the route of infection are however, unknown but it is believed to be by inhalation through the nasopharynx or by hematogenous spread to the brain (https://en.wikipedia.org/wiki/Sappinia)

Naegleria fowleri

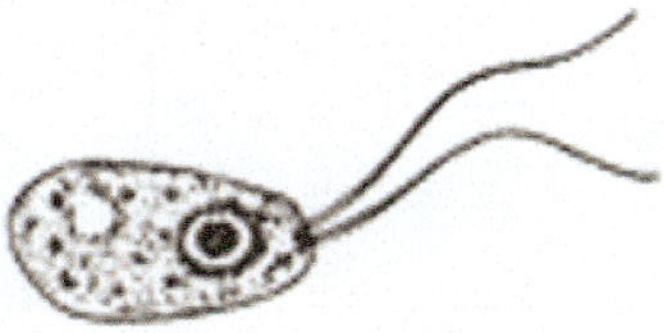

Image credit: Wikipedia

Family: Vahlkampfiidae

Common name: Brain-eating amoeba

Global distribution: It is commonly found throughout the world

Ecology: This water-based (opportunistic) species is commonly found in warm freshwater bodies such as streams, lakes, rivers, ponds and canals. It is an amphizoic organism i.e. it has the ability to exist both as a free-living ameba and as a parasitic pathogen

Description/ Biology: It is a microscopic, single-celled amoeba. There are three distinct morphological stages in its life cycle and they are : trophozoite, flagellate, and cyst . Among these stages, the trophozoite is the causative agent for infection.

Harmful organs and harm: This species has been reported to cause a rare infection of the brain called "Primary Amebic Meningoencephalitis (PAM)" , a rapidly fatal disease associated with the inflammation of the brain . This species commonly infects immunocompetent children and immunocompromised adults who go for swimming in freshwater bodies like amoeba-contaminated lakes and rivers. Infection normally occurs when this pathogen enters the nasal passages, attaches to the olfactory mucosa, and migrates to the olfactory bulbs of the brain where the ameba divides rapidly and death occurs in the victim within 7 to 10 days. This pathogen is believed to cause a fulminating hemorrhagic necrosis of the brain (Ashbolt ,2015; https://www.cdc.gov/parasites/naegleria/index.html). Symptoms associated with PAM include frontal headache, fever, nausea, vomiting, and rhinitis which are are normally followed by stiff neck and diplopia leading to coma and death (Güémez and Garcia, 2021)..

This species has also been reported to serve as an environmental host of amoeba-resisting bacterial pathogens such as NTM (Non-tuberculous mycobacteria) and legionellae. The species *Legionella pneumophila* is known to cause a pneumonia-type illness called "Legionnaires' disease " and a flu-like illness called "Pontiac fever" (Rodriguez-Zaragoza,1994; Ashbolt , 2015; Wannasan et al., 2013)

Vahlkampfia avara

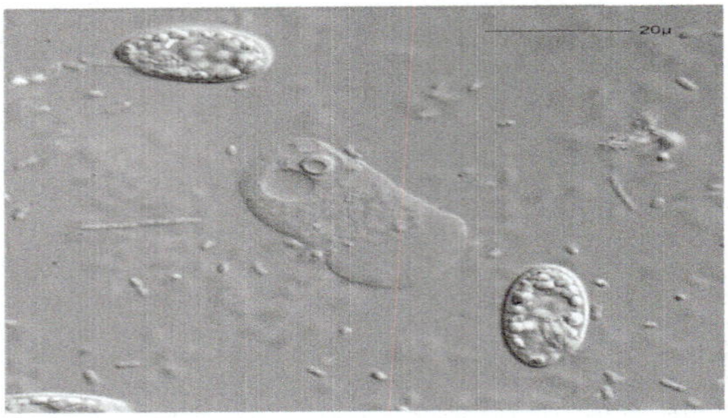

Vahlkampfia **sp.**
Image credit: Picturepest, Wikipedia

Family Vahlkampfiidae

Global distribution: Common in Iran and Mexico

Habitat: It occurs in all freshwater habitats including streams. This species has been observed in a mixed infection with *Acanthamoeba* spp.

Description: The locomotive trophozoites of this species are rounded with 10-15 µm in diameter and are long with a length range of 15-33 µm. Pseudopods of this stage are clear and eruptive. Nucleus is spherical with a size of 3-6 µm. Cysts are of 7-14 µm in diameter.

Harmful organs and Harm: The different species of this genus behave as environmental hosts of amoeba-resisting bacterial pathogens such as NTM (Nontuberculous mycobacteria) and legionellae. The species *Legionella pneumophila* has been reported to cause a pneumonia-type illness called "Legionnaires' disease" and a flu-like illness called "Pontiac fever" (Wikipedia; Rodriguez-Zaragoza,1994; Ashbolt , 2015; Wannasan et al., 2013).

The species of this genus have also been found to cause Acanthamoeba keratitis (a serious eye infection primarily affecting contact lens users) via drinking water (Ashbolt , 2015)

Vahlkampfia limax

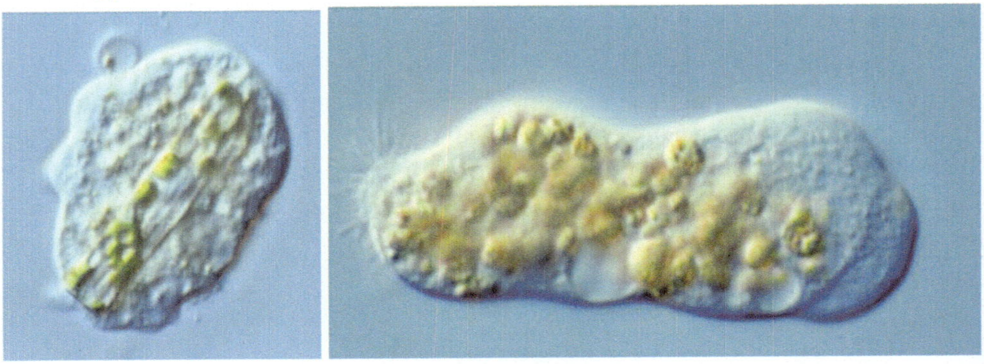

Image credit: Dr. Antonio Guillén (Reproduced with permission)

Family Vahlkampfiidae

Global distribution: Spain

Habitat: It is a water-based (opportunistic) protozoan (Ashbolt,2015) and is found in all freshwater bodies including rivers and lakes where it often lives on sediments (https://www.biodiversidadvirtual.org/micro/Vahlkampfia-limax-img1049.html).

Description: The locomotive stage of this amoeba has only one nucleus with a single nucleolus and numerous vacuoles. One broad pseudopodium is present. Small granulations are seen in the cytoplasm. It is known for its snail-like movement. The cyst stage is with a perforated wall which is consisting of two layers viz. an inner very finely fibrous layer and an outer layer with a coarser texture.

Harmful organs and Harm: As for *Vahlkampfia avara*

Vannella spp.

Family: Vannellidae

Common name: Limax amoebae

Global distribution: Spain and Mexico

Habitat: All freshwater bodies including swimming pools and streams (Ariza et al., 1989). It is a water-based (opportunistic) protozoan (Ashbolt,2015)

Description : The pathogenic species of this genus viz. *Vannella. simplex, V. platypodia, V. mira and V. miroides* possess both resting forms and locomotive trophozoites ; and the latter occur as floating and locomotive forms (Ariza et al., 1989). .

Vannella platypodia

Trophozoites of this species longer than wider and spatulate. Length and width of the forms varied continuously. Body is fan-shaped in locomotion. Length,10-25 µm; and dimeter of nucleus, 3.4-4.8 µm.

Vannella simplex

Locomotive forms of this species vary in size and are semicircular or fan-shaped; and are often with a long 'tail'. Rounded vesicular nucleus is 6—11 um in diameter and is with a single spherical central nucleolus. Floating form is of radial type and is with a number of fine, tapering hyaline pseudopodia. Length of locomotive forms , 25—80 µm and diameter of nucleus, 6—11 µm.

Vannella mira

In locomotion, this species is fan-shaped and semicircular or (rarely) spatulate. Prominent folds or lobes are seen on the frontal hyaloplasm. Single vesicular nucleus which is 6 µm in diameter and is with a central spherical endosome of 3 µm diameter. Floating form is non-symmetrical, and is with 11 straight or coiled pseudopodia,

Pseudopodia of the floating form have rounded, and are relatively thick.Length and breadth of the locomotive form, 15-35 µm; and diameter of nucleus, 4.8- 8.3 µM

Vannella miroides

Locomotive trophozoite has a fan like structure and is flattened. Length, 25-35 µM; and diameter of nucleus, 4-4.5 µM.

Harmful organs and Harm: The different species of this genus behave as environmental hosts of amoeba-resisting bacterial pathogens such as NTM (Nontuberculous mycobacteria) and legionellae. The species *Legionella pneumophila* has been reported to cause a pneumonia-type illness called "Legionnaires' disease" and a flu-like illness called "Pontiac fever" (Wikipedia; Rodriguez-Zaragoza,1994; Ashbolt , 2015; Wannasan et al., 2013).

Some of these free-living amoebae may be pathogens in their own right, such as those causing Acanthamoeba keratitis via drinking water, a serious eye infection primarily affecting contact lens users, although several cases have involved other genera (Vahlkampfia, Vannella, and Hartmannella spp.) (Ashbolt , 2015)

4.2. Sponges

Drulia ctenosclera

Image credit: Cecilia Volkmer-Ribeiro (Reproduced with permission)

Phylum: Porifera Class: Demospongiae

Order: Spongillida Family: Metaniidae

Common name: Freshwater sponge

Global distribution: Thoughout the world; it is endemic to the Amazon and Tocantins basins; Xingu River, Brazil;

Ecology: It encrusts rocky bottoms which are influenced by high water velocity and turbulence

Description/ Biology: This sponge is stony hard and is gray to black, in colour. It has an extremely hispid surface due to the poignant projections of its oscular apertures. Its skeletons are composed of silicious and calcareous spicules.

Harmful organs and harm: Harmful organs and harm: The freshwater sponges *Drulia ctenosclera* and *D. uruguayensis* are harmful to the collectors. These species cause irritating skin disorders similar to those caused by marine sponges for sponge collectors whose skin and mucous membranes may come into contact with the spicules of sponges. Lesions caused by the spicules of these sponges are disseminated, and are manifested as pruritic erythematous papules surmounted by vesicles. These lesions may also present exulcerations, crusts and secondary infections. All skin and mucous membranes can come into contact with the spicules of sponges, which occurs in the baths in pools and rivers (Haddad Jr ., 2013).

Drulia uruguayensis

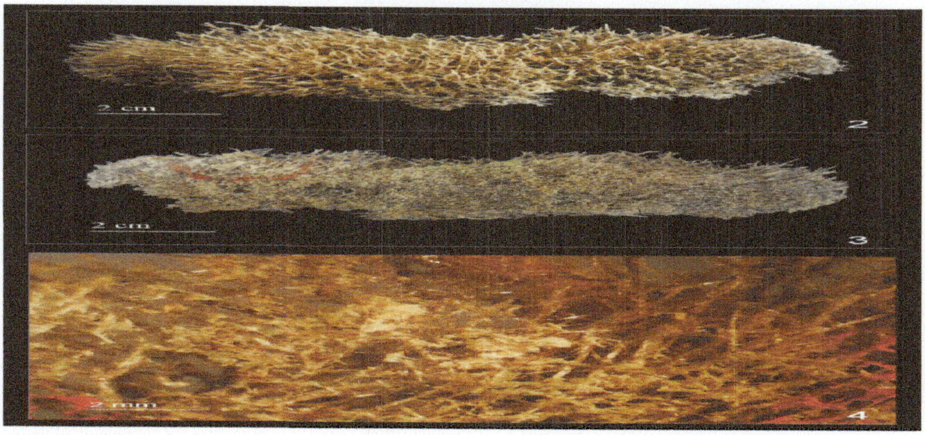

Image credit: Cecilia Volkmer-Ribeiro (Reproduced with permission)

Common name: Freshwater sponge

Global distribution: Thoughout the world; Tapajós River, Brazil; Suriname (in northeastern Atlantic coast of South America)

Ecology: Freshwater sponges are found in all freshwater bodies such as lakes and tributaries of rivers. This species encrusts rocky bottoms which are influenced by high water velocity and turbulence

Description/ Biology: Skeletons of this sponge are composed of silicious spicules

Harmful organs and harm: The silicious spicules (gemmoscleres) of both *Drulia uruguayensis* and *D. ctenosclera* have been reported to be the causative agents for intraocular lesions and ocular pathology (Volkmer-Ribeiro et al., 2006)

Ephydatia sp.

Image credit: Ond ej Macha (Reproduced with permission)

Order: Haplosclerida Family: Spongillidae

Common name: Freshwater sponge

Global distribution: Australia

Ecology: It occurs both in brackish and freshwater ecosystems

Harmful organs and harm: This unidentified species has been reported to cause irritation upon handling and unspecified substances from this sponge caused diarrhea, respiratory distress and death in animals (Hayes,2007)

Spongilla lacustris

Image credit: Wikipedia

Family : Spongillidae

Common name: Freshwater sponge

Global distribution: It is found distributed from North America to Europe and Asia; most common wide spread freshwater sponge in central Europe and Northern Britain,

Ecology: It is a common freshwater sponge lakes and canals. It can be found under rocks and logs; and can also grow on reed.

Description/ Biology: Its color is mostly white but it may have a greenish color because of symbiotic algae. Surface of the sponge is uneven and roughened by small spicules.

Harmful organs and harm: This species may cause irritation upon handling and its unspecified substances may cause diarrhea, respiratory distress and death in animals (Hayes,2007)

4.3. Trematodes

Clonorchis sinensis

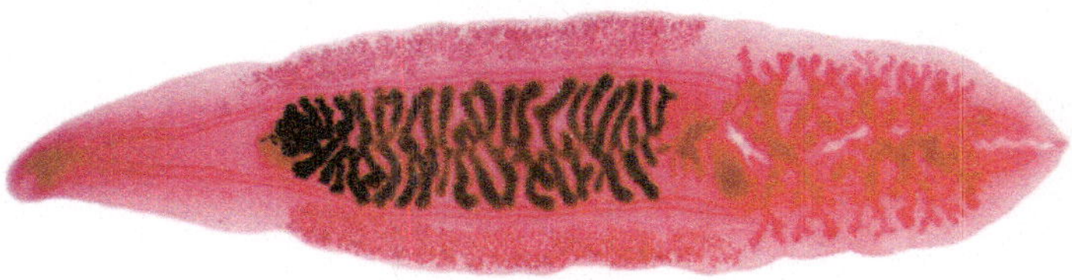

Image credit: Banchob Sripa, Wikipedia

Phylum: Platyhelminthes Class: Trematoda Family: Opisthorchiidae

Common name: Chinese or oriental liver fluke

Global distribution: It is endemic to Asia and Russia

Ecology: Its life cycle takes place in an aquatic habitat.

Description/ Biology: Adult fluke is dorsoventrally flattened and leaf-shaped. Body is slightly elongated and slender, measuring about 20 mm in length and 4 mm in width. It narrows at the anterior region into a small opening called oral sucker, which serves as the mouth. From the mouth, two tubes (caeca) which run the entire length of body are the digestive and excretory tracts. Posterior end of the fluke is broad and blunt. A poorly developed ventral sucker is lying behind the oral sucker, at about one-fourth of the body length from the anterior end. The first intermediate host of this parasite is a freshwater snail, *Parafossarulus manchouricus* or *Bulinus* sp.; freshwater fish as the second intermediate host, and mammals are the definitive hosts.

Harmful organs and harm: Humans are infected by this parasite by eating raw or undercooked fish, crabs, or crayfish from areas where this parasite is found. This fluke has been reported to infect the liver, gallbladder, and bile duct in humans.

The infection of this parasite is called "clonorchiasis" which generally appears as jaundice, indigestion, biliary inflammation, bile duct obstruction, and even liver cirrhosis, cholangiocarcinoma, and hepatic carcinoma (https://www.cdc.gov/parasites/clonorchis/index.html).

Opisthorchis viverrine

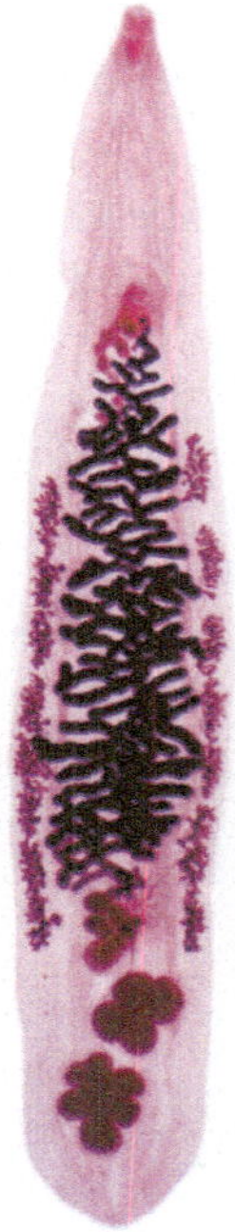

Image credit: Banchob Sripa, Wikipedia

Family: Opisthorchiidae

Common name: Southeast Asian liver fluke

Global distribution: It is commonly found in Thailand, the Laos, Vietnam, and Cambodia.

Ecology: It occurs in freshwater habitats with stagnant or slow-moving waters including ponds, river, aquaculture, swamps, rice fields

Description/ Biology: Adult fluke is transparent, dorsoventrally flattened like a leaf and is shaped like a lancet. A typical individual is 7 mm long and 1.5 mm wide. Anterior end is more pointed and is a mouth-like structure called oral sucker. Just behind the oral sucker is a similar structure called ventral sucker. These suckers are the organs of attachment. It is completing its lifecycle in three different categories of animals which serve as intermediate hosts or definitive hosts. Snails of the genus Bithynia are the first intermediate hosts; cyprinid fishes are the second intermediate host; and the definitive hosts include humans and other mammals such as dogs, cats, rats, and pigs.

Harmful organs and harm: People are infected after eating raw or undercooked fish. Infection with the parasite is called "opisthorchiasis". Infection of this parasite also increases the risk of cholangiocarcinoma, a cancer of the bile ducts (Wikipedia).

Most infections due to this fluke are asymptomatic. Most pathologic manifestations of this fluke often result from inflammation and intermittent obstruction of the biliary ducts. In mild cases, manifestations of this fluke include abdominal pain, diarrhea, constipation and dyspepsia (https://www.cdc.gov/dpdx/opisthorchiasis/index.html) .

Kaewpitoon et al.(2008) reported that opisthorchiasis caused by this fluke remains a major public health problem in many parts of Southeast Asia, including Thailand, Lao PDR, Vietnam and Cambodia.

4.4. Nematodes

Dracunculus medinensis

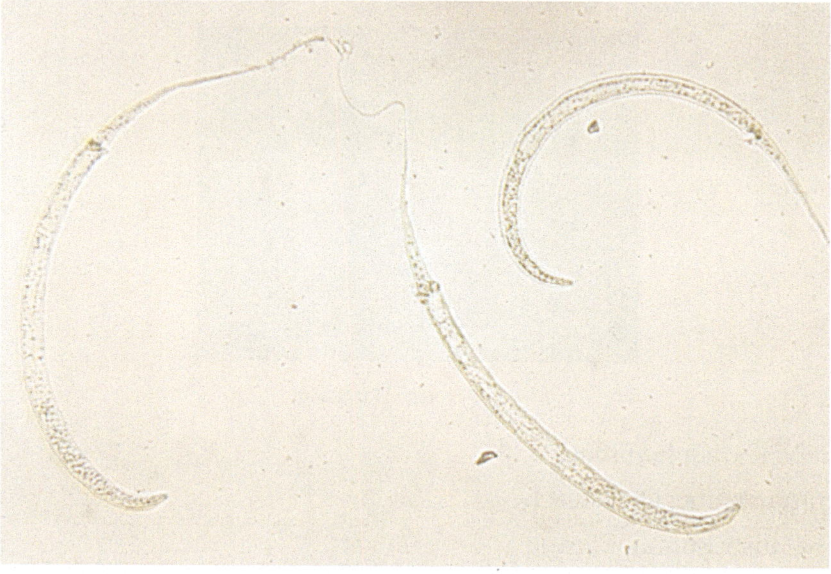

Photo Credit:Content Providers(s): CDC/Dr. Mae Melvin , Wikipedia

Phylum: Nematoda Class: Chromadorea Order: Rhabditida

Family: Dracunculidae

Common name: Guinea worm

Global distribution: Chad, Angola and South Sudan

Ecology: Its larvae are found in fresh water, where they are normally ingested by copepods of the genus *Cyclops*

Description/ Biology: While the male worm measures 4 cm, its (disease causing) female grows up to 80 cm. Both male and female worms have a small, triangular mouth which is surrounded by a quadrangular, sclerotized plate. No lips are present. During its life cycle, the larvae of this species develop within the copepod into an infective stage. When this infected copepod is ingested by a mammalian host via contaminated drinking water, this copepod gets dissolved by stomach acid and dies and the *Dracunculus medinensis* larvae are subsequently released and migrate through the wall of the mammalian intestine, and enter the abdominal cavity where they mature into adults.

Harmful organs and harm: This species causes dracunculiasis, also known as Guinea worm disease (GWD). Eberhard et al.(2016) reported on the occurrence of a third-stage (infective) larva of this species and this larva which is believed to be the causative agent of GWD was recovered from a frog of this species, *Phrynobatrachus francisci* from Chad.

4.5. Annelids

Theromyzon tessulatum (= *Hirudo tessulata*)

Image credit: Wikidata

Family: Glossiphoniidae

Common name: Duck leech

Global distribution: Europe

Ecology: It normally occurs in freshwater reservoirs. It however, prefers stagnant water bodies

Description/ Biology: Cuticle of this species is transparent, olive green or brownish in colour, Larger individuals are often exhibiting six longitudinal rows of small yellow spots. This leech is known for frequently changing its shape and it grows up to 6 cm long when extended. Its posterior sucker is broad and conspicuous while the anterior sucker is small.

Harmful organs and harm: It is a blood sucking leech. Leech bite is also a danger to humans. It is reported that the presence of hirudin in the wound following a leech bite may cause oozing to continue for several hours. Further, gut bacteria can cause wound infection. There may also be a delayed irritation and itching after a bite. Furthermore, allergy to leech bite has also been reported (https://australianmuseum.net.au/learn/animals/worms/leeches/)

Hirudo medicinalis

Image credit: Karl Ragnar Gjertsen Please credit Karl Ragnar Gjertsen, Wikipedia

Phylum: Annelida Class: Clitellata Family: Hirudinidae

Common name: European medicinal leech

Global distribution: Central and northern Europe

Ecology: It has an amphibious mode of life; needs both land and water; and resides exclusively in fresh water. A typical habitat for this species is a small pond with a muddy bottom edged with reeds and where frogs are at least seasonally abundant

Description/ Biology: Medicinal leech has a cylindrical, dorsoventrally flattened body which is 10–15 cm in length and is divided into 33-34 segments. Dorsal side is dark brown to black, bearing six longitudinal, reddish or brown stripes, and its ventral surface is speckled. A posterior and anterior disk-shaped suckers are present. Anterior sucker surrounds the oral opening where the teeth for incison are located. In addition, this leech has five pairs of eyes located at its front end

Harmful organs and harm: Medicinal leech is parasitic on humans and is a source of unpleasant emotions for leech victims and bystanders alike.

Medicinal leech has been reported to ingest 3–10 times its body weight in blood at a single feeding. It sucks blood without causing pain, by dissecting the skin of the prey with its 3 sharp teeth located in its upper jaw, simultaneously releasing mediators in to the bite via its saliva. This leech may cause external skin bleeding, ecchymosis and various skin lesions . Patients with leech bite may have itchy lesions on the calves, knees, and feet.

The cream (Hirucrème) made with the crushed medicinal leech has been reported to cause contact eczema (contact dermatitis) (Blaise et al., 2002)

Haddad Jr. (2013) reported that this species may be harmful in the following lines:

a. it may cause deaths due to uncontrollable bleeding by the action of hirudin;

b. it removes considerable quantities of blood that do not compromise blood clotting;

c. it may cause severe infections by *Mycobacterium marinum*; and

d. it may cause acute anemia in patients who are often parasitized.

Hirudo verbana

a-Dorsal view; b- Ventral view
Image credit: Wikiceylan , Wikipedia

Common name: Medicinalis Bloodsucking Leech ; Hungarian Leech

Global distribution: Switzerland, Austria, Germany, Italy, Slovenia, Croatia, Bosnia , Herzegovina, Serbia, Montenegro, Macedonia, Greece, Hungary, Moldova,

Ukraine, Russian Federation, Turkey , Uzbekistan, Romani, Bulgaria and southern Slovakia

Ecology: It is common by seen in rivers

Description/ Biology: Dorsal side of the body of this species has a dark appearance with orange paramedian dorsal stripes. Venter is unicoloured and greenish , with a pair of black ventrolateral stripes

Harmful organs and harm: The jaws and the body surface of this species have been reported to harbour fungal pathogens such as *Candida albicans, C. ciferrii, C. krusei, C. tropicalis, Trichosporon asahii and T. asteroides*. The isolation of a high number of yeast-like fungal strains from this suggests that this leech may serve as a vector of potentially pathogenic human fungal species (Litwinowicz and Blaszkowska, 2013)

Limnatis nilotica

Image credit: Francesco Saverio Monticelli and Monricelli, F. S. , Wikipedia

Family: Hirudinidae

Common name: Horse leech or Nile leech

Global distribution: Southern Europe, North Africa, and the Middle East,

Ecology: It inhabits stagnant ponds, streams and shallow lakes.

Description/ Biology: This species grows to a maximum length of 12 cm . It has powerful jaws and a sucker at both the anterior and the posterior end. Body colour is dark green and there are green spots in rows on the dorsal surface and bands of yellowish-orange and green on its sides.

Harmful organs and harm: Smaller sized leeches of this species have been reported to enter the human body. When these leeches are ingested with drinking water, they may first attach themselves to the linings of the nose or throat and then enter the lungs. Some leeches are also likely to enter the excretory openings of humans who bathe in infested waters (https://www.britannica.com/animal/leech#ref57162).

Macrobdella decora

Image credit: Smithsonian Environmental Research Center, Wikipedia

Family: Hirudinidae

Common name: North American medicinal leech

Global distribution: It is native to Canada and the northern U.S.A. It is found distributed from Southern Canada to the southern United States

Ecology: It is found in shallow, protected waters

Description/ Biology: It grows up to 10 cm long and 1 cm wide. Its upper surface of the body is olive with a row of orange spots down the middle.

Harmful organs and harm: It is the only hematophagic, or blood-sucking, freshwater leech (https://askinglot.com/are-freshwater-leeches-harmful). It has been reported to take human blood regularly (https://www.des.nh.gov/organization/commissioner/pip/factsheets/bb/documents/bb-10.pdf)

Macrobdella mimicus

Image credit: Robert Aguilar, Smithsonian Environmental Research Center.(Reproduced with permission)

Family: Hirudinidae

Common name: North American medicinal (human-feeding) leech

Global distribution: Eastern United Stat; from Georgia to New York

Ecology: Freshwater wetlands

Description/ Biology: It is an orange-spotted, olive-green leech. It has multiple reproductive pores along the bottom of its body , known as gonopores and accessory pores, which are located in a different position

Harmful organs and harm: When these leeches are disturbed, they can regurgitate the bacteria which sit in their intestines to facilitate their digestion into the wound, leading to infection (https://www.kuow.org/stories/a-new-bloodsucking-leech-species-found-hiding-outside-washington-d-c).

Limnodrilus hoffmeisteri

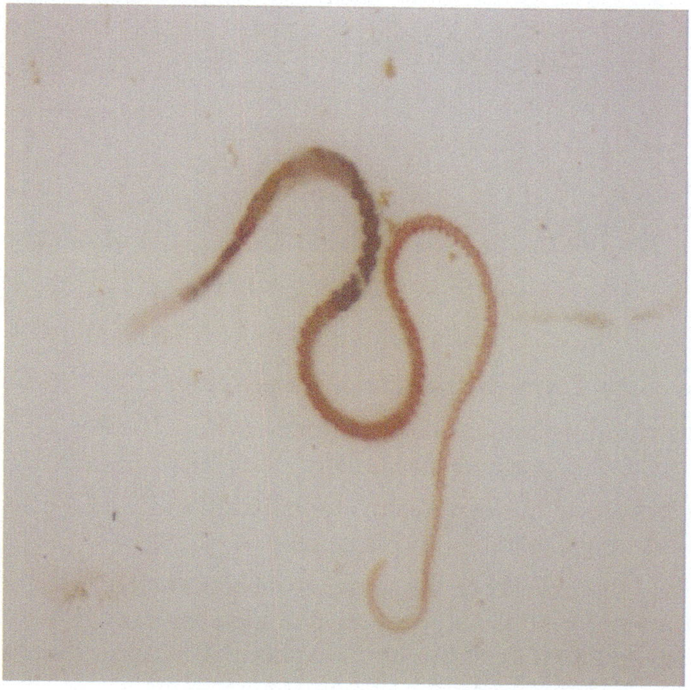

Image credit : Fengjiao Peng , Wikipedia

Family: Naididae

Common name: Red worm

Global distribution: Throughout world

Ecology: It lives in sandy and coarse sediments of watercourses, pond and lakes.

Description/ Biology: It has a cylindrical body with 55-95 segments and it grows to a maximum length of 40 mm. A conical shaped head without eyespots is present. Each of the segments of this species can regenerate into a new individual when it is separated from the worm body. On each side of the segment, there is an upper and lower bundle of setae.

Harmful organs and harm: This species has been reported to cause nose bleed, itching, sneezing, and rhinorrhea when it enters the nasal cavity of the infected people (Liu et al., 2017)

4.6. Arthropods

Argyroneta aquatica

Image credit: Baupi, Wikipedia

Phylum: Arthropoda Class: Arachnida Family: Cybaeidae

Common name: Diving bell spider or water spider

Global distribution: British Isles and northern Asia ;Turkey, Caucasus, Russia, Iran, Central Asia, China, Korea and Japan

Ecology: It is found in clean freshwater habitats with aquatic vegetation, such as lakes, ponds, canals, marshes and streams

Description/ Biology: When the spider is out of the water, its colour ranges from mid to dark brown, albeit its hairs on the abdomen are dark grey with velvet-like appearance. Males are more active and are about 30% larger than females. While males measure 10–15 mm in head-and-body length females are only 8–12 mm.

Harmful organs and harm: The bite of this species is quite painful to humans and it may cause localised inflammation, vomiting, and feverishness which may however, disappear within a few days of the bite (https://www.scienceabc.com/nature/animals/diving-bell-spider-anatomy-habitat-bite-facts.html).

Lethocerus delpontei (= *Belostoma delpontei*)

Image courtesy: Wikipedia

Class: Insecta Order: Hemiptera Family: Belostomatidae

Common name: Giant water bug

Global distribution: Tropical, subtropical and temperate areas of the world.

Ecology: It is found in freshwater environments

Description/ Biology: This bug reaches a length of about 10 cm. When the animal is totally submerged beneath the surface of the water, it respires by means of air stores which adhere to its body and communicate with the spiracles. Its mesothoracic spiracle is situated between the prothorax and mesothorax; metathoracic and first abdominal spiracles lie posteriorly in the mesothorax and metathorax, respectively. The mesothoracic and metathoracic spiracles are of less respiratory importance.

Harmful organs and harm: The toxic saliva of this species can cause painful accidents in humans. The bite inflicted by its proboscis provokes severe pain and reversible paralysis of limbs can occur (Haddad, Jr., 2013). Walker et al. (2016) reported that the bite of this species may paralyse frogs within several minutes. These venomous animals can also cause: skin necrosis; erysipelas after bites; and granuloma annulare in humans.

4.7. Molluscs

Radix natalensis (= *Lymnaea natalensis*)

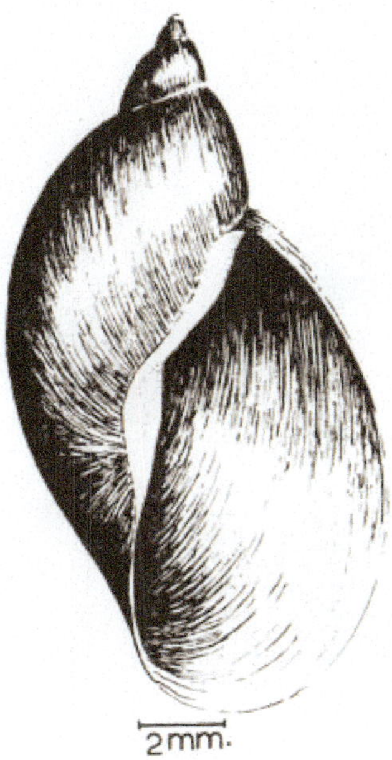

Image credit: Wikipedia

Phylum: Mollusca Class: Gastropoda

Subclass: Pulmonata Family: Lymnaeidae

Common name: Not designated

Global distribution: This species occurs commonly in Africa; from northern Africa to southern Africa

Ecology: It lives in shallow, permanent freshwater bodies in depths of 0–4 cm. It prefers plant detritus as a substrate.

Description/ Biology: Shape of the shell of this species is elongate ovoid. Shell is colorless or light brown in color. Height of its aperture covers about 3/4 of the shell height. Lip of the aperture is thin and sharp. Umbilicus is closed. Width of the shell is about 7 mm. and maximum height of the shell is 19 mm (Wikipedia)

Harmful organs and harm: It is a major intermediate host for the trematode parasite, *Fasciola gigantica* (Wikipedia). "Fasciolosis" which is caused by *Fasciola* spp. is a serious infectious parasitic disease infecting humans and animals worldwide. Ibrahim and Ahmed (2019) reported that the infected snails included five types of the cercarial stages beside *Fasciola* cercariae and three types of metacercariae. These cercariae were the xiphidio cercariae, gymnocephalus, echinostome cercariae, virgulate cercariae, echinochasmus pelecani, and palaeorchis cercariae.

Biomphalaria amazonica

Image credit: Dr Jean-Pierre Pointier (Reproduced with permission)

Family: Planorbidae

Common name: South American snail

Global distribution: North and Centre-West of Brazil and North of Bolivia

Ecology: It is common in ponds and rivers

Description/ Biology: As in all planorbids , shell of this species, is sinistral in coiling, but it appears to be dextral as is carried upside down by the animal.

Harmful organs and harm: It is a potential intermediate host of *Schistosoma mansoni* (Platyhelminthes: Trematoda) and a vector of "schistosomiasis" also known as snail fever, is associated with malaria and yellow fever which are caused by parasitic flatworms of the genus Schistosoma. Schistosomiasis is a neglected tropical disease affecting over 200 million people worldwide. In affected humans, the urinary tract or the intestines may be infected. Symptoms observed include abdominal pain, diarrhea, bloody stool, or blood in the urine (Fernandez and Thiengo,2006; Pointier et al., 1966)

Biomphalaria glabrata

Image credit: Ponder ; Photo Anders Hallan, Australian Museum (Reproduced with permission)

Common name: Marsh snail

Global distribution: Africa, North America, South America

Ecology: It is predominant in irrigation systems; small streams, ponds and marshes.

Description/ Biology: As in all planorbids , shell of this species is sinistral i.e. left-handed coiling.

Harmful organs and harm: vide *Biomphalaria amazonica*

Biomphalaria pfeifferi

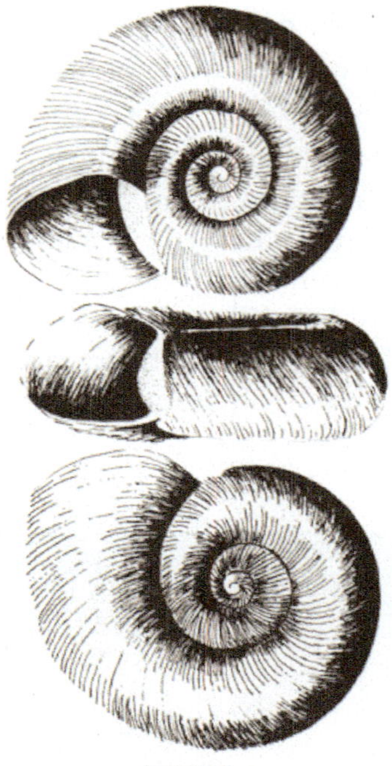

Image credit: G. Oberholzer; J. A. Van Eeden - Oberholzer G.; Van Eeden J. A. , Wikipedia

Common name: Bloodfluke planorb (Freshwater snail)

Global distribution: Western Africa

Ecology: This species occurs in a wide variety of freshwater-bodies, but the highest percentages were reported from dams, rivers and man-made water channels.

Description/ Biology: Shell of this species is discoid. Otarigho et al. (2015) reported on the morphological characters of the shells of this species collected from five water bodies across south-western Nigeria and the values are shell height , 2.8; shell width , 8.7; aperture height , 3.0; aperture width , 3.1; spiral length , 2.5; and aperture circumference , 8.1.

Harmful organs and harm: This snail is a medically important pest, as it transfers the disease schistosomiasis. It is the most important intermediate host in tropical Africa for the parasite, *Schistosoma mansoni* which is causing intestinal bilharzia in man. Alharbi et al. (2019) reported that colonization of this species in Lake Malawi (Tanzania) is of great concern, because active *Schistosom mansoni* infections were largely found in local children. This finding highlighted the emergence of intestinal schistosomiasis in the affected areas.

Biomphalaria prona

Image credit: Molluscan Base (CC)

Common name: Not reported

Global distribution: This species is restricted to Lake Valencia (Venezuela)

Ecology: It is common in freshwater lakes. It has also been found attached to the aquatic plants such as *Potamogeton* sp.

Description/ Biology: Shell of this species has 3 ½ whorls which are rapidly increasing in diameter. Its outer whorl is expanded and deflected leftward at the apertural region and right side is deeply umbilicate. Largest specimen measures 7.5 mm in diameter.

Harmful organs and harm: vide *Biomphalaria amazonica*

Biomphalaria straminea

Image credit: Guido Poppe and Philippe Poppe (Reproduced with permission)

Common name: Air-breathing freshwater snail, ram's horn snail

Global distribution: Southeastern South America; Caribbean; north- east of Brazil, Hong Kong

Ecology: It is common in natural water sources.

Description/ Biology: Shell of this species, like all planorbids is sinistral in coiling, but is carried upside down and thus appears to be dextral.

Harmful organs and harm: vide *Biomphalaria amazonica*

Biomphalaria sudanica

Image credit: Guido Poppe and Philippe Poppe (Reproduced with permission)

Global distribution: East Africa

Common name: No common names for this taxon

Ecology: It is commonly occurring in shallow water near the shoreline in Lake; and in swamp habitats. It is highly adapted to use the oxygen content of its habitats.

Description/ Biology: Shell of this species, like all planorbids is sinistral in coiling, but it is carried upside down and thereby the shell appears to be dextral.

Harmful organs and harm: This species is an intermediate host for not only the common parasite *viz. Schistosoma mansoni* but also for *S. rodhaini* (https://en.wikipedia.org/wiki/Biomphalaria_sudanica) (vide *Biomphalaria amazonica* also).

Biomphalaria tenagophila

Image credit: Gábor Majoros, Wikipedia

Common name: No common names for this taxon

Global distribution: It is a neotropical species and it is a native species of Peru, Brazil, Uruguay and Argentina.

Ecology: It is common in standing water or freshwater bodies

Description/ Biology: Shell of this species is sinistrally coiled (left-handed coiling). Its flat shells are yellow-brown and discoidal with 6 increasing whorls. Last whorl is rounded and intermediate whorls are slightly angled on the left side. Aperture is circular or slightly ovate and is angled toward the left side of the shell . Outer surface of the whorls are with fine, parallel, rib-like transverse lines . Maximum width of the shell is 21 mm and height ,6.5mm. Largest shell may have 6.5 whorls.

Harmful organs and harm: This species is a medically important pest, as it transfers the intestinal schistosomiasis disease (Caldeira et al., 2016) .

Bulinus globosus

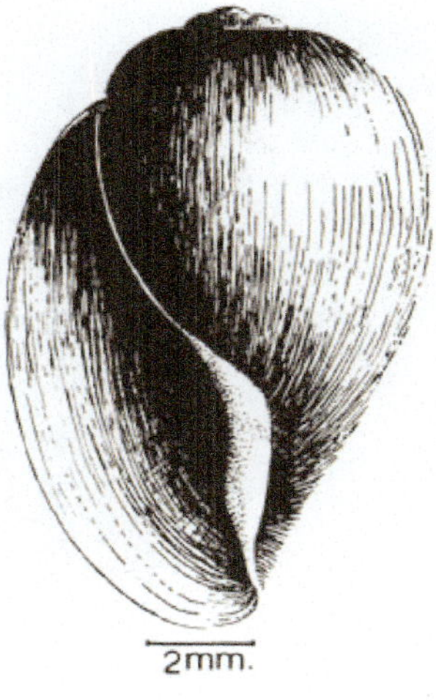

Image credit: G. Oberholzer & J. A. Van Eeden - Oberholzer G. & Van Eeden J. A. , Wikipedia

Family Planorbidae

Common name: No common names associated with this taxon

Global distribution: Lake Victoria, one of the African Great Lakes.

Ecology: It inhabits patchy and transient water bodies prone to large seasonal variations in water availability (Nyakaana et al., 2013),

Description/ Biology: Shell of the species of this is sinistral has a very large body whorl and a small spire..Snails collected from Eko-Endo Lake (Nigeria) showed the highest shell height (10.6 mm), shell width (6.5mm), aperture height (7.7mm) and aperture width (4.2mm). On the other hand, shells of Orori River (Africa) had the highest spiral length (2.6mm) (Falade and Otarigho, 2015).

Harmful organs and harm: This species has been reported to be a key intermediate host for *Schistosoma haematobium* that causes urinary schistosomiasis (Nyakaana et al., 2013). Water temperature had a non-linear effect on snail fecundity and snail growth. Snails maintained at 15.5 °C and 36.0 °C did not produce egg masses . On the other hand , snails maintained at 25.8 °C laid 344 and 105 more egg masses than snails at 31.0 °C and 21.2 °C, respectively. Water temperature therefore influences the fecundity, growth, survival and parasite development in the snail and thus decides the time it takes the parasite to complete the life cycle. This temperature variations have been reported to show implications on transmission of schistosomiasis in the context of global warming (Kalinda et al., 2017).

Bulinus truncates

Image credit: Fred A. Lewis, Yung-san Liang, Nithya Raghavan & Matty Knight - Fred A. Lewis, Yung-san Liang, Nithya Raghavan & Matty Knight. The NIH-NIAID Schistosomiasis Resource Center. Wikipedia

Common name: Air-breathing freshwater snail, Ramshorn Snail

Global distribution: Africa: Egypt, Marocco Northern Sáhara, D.R. Congo, Malawi and Ethiopía

Ecology: it is commonly found in irrigation canals

Description/ Biology: Shell of the species of this is sinistral has a very large body whorl and a small spire. Two populations of this species collected from Turkey showed the average values of shell height, shell width, aperture height and aperture width as 8.1mm, 5.7 mm, 4.5 mm and 3.1 mm respectively (Sesen, 2004)

Harmful organs and harm: This species is an intermediate host for *Schistosoma haematobium* and *Paramphistomum cervi* and *Paramphistomum cervi* and *P. microbothrium.*

Afrogyrorbis natalensis (= *Ceratophallus natalensis; Planorbis natalensis*)

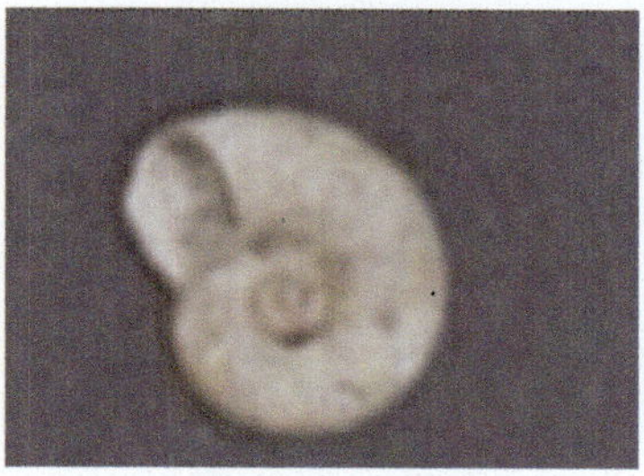

Image credit: Wikiwand

Family: Planorbidae

Common name: Freshwater snail

Global distribution: Western Indian Ocean and eastern and southern Africa

Ecology: It is often found in ephemeral rain pools of cooler areas and in marshes.

Description/ Biology: It has sinistral or left-coiling shell. It is reported that the microspinous penis of this species may be adapted to the function of sperm displacement (Brown, 2001).

Harmful organs and harm: It can act as intermediate host of a variety of helminth parasites elsewhere in Africa,

CHAPTER 5
HARMFUL FRESHWATER VERTEBRATES

5.1. Fishes

5.1.1. Cartilaginous Fishes

Carcharhinus leucas

Image credit: Chaloklum Diving , Wikipedia

Subclass: Elasmobranchii Order: Carcharhiniformes Family: Carcharhinidae

Common name: Bull shark, freshwater blue shark

Global distribution: It is commonly found worldwide. Its population is great especially in United States of America.

Ecology: It is a freshwater and coastal shark shark inhabiting shallow waters such as bays, rivers, lakes and estuaries. It is capable of penetrating far up rivers and hypersaline bays and even littoral lagoons

Description/ Biology: It has a short, broad and blunt snout and small eyes.

Adults are a light to dark gray on their dorsal side and white on the ventral side. Teeth are a broad jagged triangle up top and a thin jagged triangle along the jawline. Its placoid scales are sharp and overlapping, and they look like pointed triangles. Adult females which are larger than males are 242 cm in total length and are weighing 111kg. Males weigh 95kg and are 228cm in length (https://animaldiversity.org/accounts/Carcharhinus_leucas/)

Harmful organs and harm: It is a traumatogenic fish. It may be more dangerous to humans than any other species of shark (Wikipedia) and they are known to attack humans (https://animaldiversity.org/accounts/Carcharhinus_leucas/)

Maculabatis gerrardi **(** = *Himantura gerrardi***)**

Image credit: Hamid Badar Osmany, Wikipedia

Order: Myliobatiformes Family: Dasyatidae

Common name: Sharpnose stingray, whitespotted whipray

Global distribution: Indo-West Pacific: India to New Guinea; Japan

Ecology: It is commonly found in coastal areas including estuaries; and it has also been reported from the Ganges River (Wikipedia)

Description/ Biology: Disc of this species is rhomboidal and its trunk is thickened; pectoral- fin apex is narrowly rounded; and tail is very slender, elongated and whip-like with a single venomous spine (sting). Mouth is narrow with a pair of enlarged central papillae on its floor. Coloration of disc is brown or grayish brown dorsally, with scattered white spots posteriorly, and white ventrally. Tail is brownish with a row of white spots dorso-laterally. It attains a maximum length of about 2m and its maximum disc width is about 90 cm.

Harmful organs and harm: Injection of the extract from the venomous spines of this species produced intense pain in all affected humans. Further, redness was observed in two patients , and spasm and seizure were each recorded in only one patient. The crude venom extract showed an LD_{50} value of about 100 mg kg–1 in experimental mice. Further, cardiac symptoms such as an increase in pulse rate and various changes in electrocardiogram (ECG) parameters such as T and Q values, and PR and RR intervals were observed in experimental rats (Dehghani et al., 2009).

Urogymnus polylepis (= *Himantura polylepis; Himantura chaophraya*)

Image credit: Nelson Annandale , Wikipedia

Common name: Giant freshwater stingray, freshwater whipray

Global distribution: Thailand, Malaysia and Island of Borneo

Ecology: It is found on the sandy bottoms in large rivers, at depths of 5 to 20 m. Females are however,found in estuaries also

Description/ Biology: This species has been reported to be the largest freshwater fish in the world and the largest among 200 species of stingrays. It reaches a weight of 600kg and 5 m in total length (https://wwf.panda.org/discover/our_focus/wildlife_practice/profiles/fish_marine/stingray/ ; https://animaldiversity.org/accounts/Himantura_chaophraya/). Body is flat. While its tail is very smooth on the dorsal side, ventral side has a spine with serrations and a venom gland. Two pelvic fins are seen on either side of the tail. Pectoral fins have about 160 pectoral radials, which are small bone-like structures. Mouth is found on the underside of the disk and its two jaws are filled with small teeth. Two parallel rows of gill slits are seen posterior to the mouth. Its dorsum is darkly colored. Based on its size , this species is believed to have a life span of about 40 yrs.

Harmful organs and harm: It has the longest venomous spine (35 cm) among its counterparts. This ray uses its spine mainly for defense and it can also be lethal to humans.

Paratrygon aiereba

Order: Myliobatiformes Family: Potamotrygonidae

Common name: Discus ray, ceja ray

Global distribution: South America: Amazon River basin and Orinoco basin

Ecology: This demersal species inhabits major river systems and it may approach the margins during the night. It however, prefers shallow areas and near banks. River stingrays cannot tolerate low temperatures and they have been reported to grow well at a temperature of 26-27°C/79-81°F (https://www.practicalfishkeeping.co.uk/features/definitive-guide-to-south-american-freshwater-rays/)

Description/Biology: Upper surfaces of these stingrays are covered with denticles (sharp tooth-like structures). Body is brown or yellow with spotted or mottled patterns. Males have modified pelvics and two claspers ; and females possess rounded pelvic fins. It has robust tails with stingers in mid-distal position. Its maximum disc width is 1.2 m and weight , 50 kg.

Harmful organs and harm: These stingrays have one or two venomous stingers which are located on the dorsal portion of the tail. Stingers which are composed by modified dermal denticles are bilaterally retroserrated, and are covered by glandular and integument tissues. These tissues are abundant in toxin-secreting cells which are responsible for the envenomation. According to available reports, these rays are not dangerous unless they are stepped on. Haddad Jr. et al. (2013) reported that this species is very unlikely to cause harm due to its short tail and the position of the spines. However , its envenomation normally produces an extremely painful injury which causes necrosis and ulceration in tissues underlying the wound, due to the action of the venom. Further, retaining fragments of stingers are likely to be retained in the wound with triggering of infectious process. In affected people , injuries may cause a temporary or permanent physical incapacity, and the injured may also develop sequelae in the affected limb. Lethal wounds may also occur in patients when stingers reach their vital organs (da Silva et al., 2020).

Plesiotrygon iwamae

Image credit: User: Davepape , Wikipedia

Common name: Long-tailed river stingray , antenna ray

Global distribution: South America: Amazon River basin; and from Ecuador to Belém.

Ecology: This demersal species inhabits the main channel of the Amazon River and lower parts of its major tributaries

Description/ Biology: It has an oval disc and is longer than wide. Eyes are very small and are not protruding from head. Spiracles are wide and rhomboidal. Mouth is also wide and teeth set in quincunx are very small and are in numerous rows. Pelvic fins are protruding well from posterior disc region. Clasper is stout and short, with a rounded posterior tip. Lateral and dorsal tail folds are absent. Caudal stings are positioned far posteriorly on dorsal tail. Dorsal color is light gray. Several large, faint, white spots or irregular ocelli are present over middisc or posterolateral disc. Larger, irregular spots are often surrounded by smaller, faint spots. Light irregular vermicular markings may also be present. Its disc width is 58 cm (de Carvalho and Ragno,2011)

Harmful organs and harm: Freshwater stingrays are known for causing injuries which are associated with intense pain and pathological changes such as oedema, erythema and necrosis at the lesion site. Apart from the injuries, stinger and dorsal extracts of this species developed coagulative necrosis of the muscle tissue, and inflammatory infiltrates such as neutrophils, macrophages, and a reduced number of eosinophils and lymphocytes. Further these extracts induced systemic rhabdomyolysis (breakdown of damaged skeletal muscle) (Lameiras et al., 2014).

Plesiotrygon nana

Image credit: Frederik Mollen, Elasmobranch Research Belgium (Reproduced with permission)

Common name: Dwarf freshwater stingray, black tailed antenna ray

Global distribution: Upper and middle Amazon River drainage

Habitat: It is common in river systems

Physical Description: Dorsal surface of the disc of this species has highly convoluted yellowish lines or small spots which are in the form of very fine rosettes or a mixture of spots and irregular ocelli. The other characteristic feature include a circular disc, broadly rounded snout, very small rhomboidal spiracles, short snout and narrow mouth and nostrils. Denticles on the dorsal surface of the tail are small, scattered, and are not in the form of a row of enlarged spines. This species is also known for its small adult size and disc length or width which is less than 25 cm.

Harmful organs and harm: As it is a new species with insignificant dorsal venomous spines, further investigations are needed on the envenomation of this species.

Potamotrygon amandae

Common name: River stingray

Global distribution: South America: Argentina, Brazil, Bolivia and Paraguay, Paraná-Paraguay basin

Ecology: Riverine systems

Description/ Biology: Its dorsal disc is more grayish or dark brown. Ocelli when present are bicolored i.e whitish, light gray or light yellow central area surrounded by a black peripheral ring. Ventral disc coloration is grayish. Dermal denticles are smaller in size.

Harmful organs and harm: Its venom gland possesses high abundance of an enzyme viz. hyaluronidase which helps in the absorption and diffusion of venom in the victim's tissues thereby increasing systemic poisoning (deOliveira Jr.et al., 2016). Hyaluronidase has also been reported to increase tumor malignancy.

Potamotrygon falkneri

Image credit: Andrew Kuchling, Wikipedia

Common name: Large spot stingray

Global distribution: South America: Argentina, Bolivia, Brazil, Peru. Paraná-Paraguay River basin, Piquiri River, La Plata River and Amazon Basin

Ecology: Riverine systems

Description/ Biology: Its dorsal disc is dark brown with yellow or white spots which may be oval, circular or vermicular. Its robust tail has 1-3 irregular rows of spines (stingers) at its mid-dorsal portion.

Harmful organs and harm: Its tail and its stingers have been reported to cause accidents frequently in humans. Haddad Jr. et al.(2013) reported that the victims are commonly stung on the lower limbs during the day while swimming. Hayes (2007) reported that the venom of this species exhibited caseinolytic, gelatinolytic and hyaluronidase enzymatic activities. Victims of this species experienced edema, skin necrosis and ulceration. Further, the freshwater stingray envenomation yielded systemic symptoms such as dizziness, fever, migraine, myalgia and vomiting.

Potamotrygon henlei

Image credit: Christine Schmidt from Laurel, USA , Wikipedia

Common name: Bigtooth river stingray, Tocantins river ray

Global distribution: South America: Tocantins and Araguaia rivers, Brazil

Ecology: It dwells in rivers and it tends to approach the margins during the night.

Description/ Biology: Its dorsal color pattern is blackish to grayish- brown with few ocelli which are rounded and yellowish to light orange in color. Further, these ocelli have thin black contours and are with light centers. Stingers of the tail are in mid-distal position

Harmful organs and harm: Haddad Jr. et al. (2013) reported that its robust tail with stingers may frequently cause accidents Both venom and mucus from the sting, and mucus from the skin of this species enhanced vascular permeability and promoted nociception in mice (Monteiro-dos-Santos et al., 2011)

Potamotrygon leopoldi

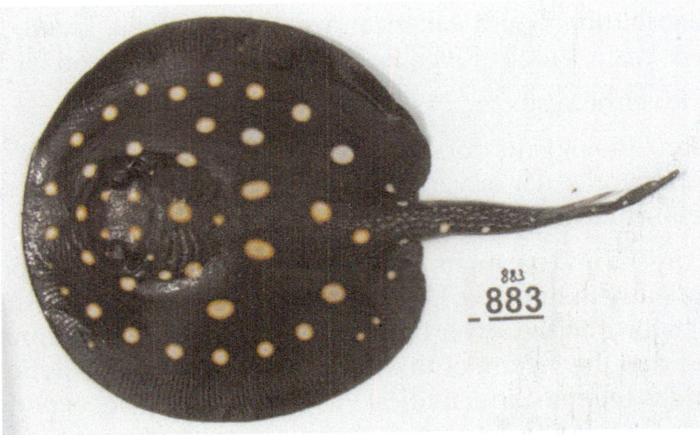

Image credit: Frederik H. Mollen, Elasmobranch Research Belgium (Reproduced with permission)

Common name: Leopold stingray, White-blotched river stingray

Global distribution: South America: Xingu and Fresco Rivers, Xingu River basin,Brazil

Ecology: Riverine systems

Description/ Biology: Its dorsal disc is with intense black color. Its ocelli which are few and large are yellow to whitish colored. Mid disc has incomplete rings or sickle shaped figures with black centers. Maximum values of length of male and female are 75 and 100 cm respectively and the corresponding disc length, 45 cm and 65 cm.

Harmful organs and harm: This species can cause severe injuries with the venomous dentine spine located at its tails. The juvenile extracts of this species induced toxic response at a venom concentration of 0.002 µg/µl and the IC_{50} occurred at a concentration of 0.01 µg/µl. On the other hand, the venom extracts of mature specimens showed the first toxic response at a dose of 0.02 µg/µl and the IC_{50} was observed at a concentration of 0.1 µg /µl (Kirchhoff et al., 2014)). This species has been designated as dangerous than other supposedly life-threatening species, such as piranha (http://www.duketownstingrays.com/Files/2/60000/60267/Attachments/Product/49zdG998465t182d7098g040k042cv16.pdf).

Potamotrygon limai

Common name: It is a new species

Global distribution: South America: Jamari River and Madeira River basin in Brazil.

Ecology: This benthopelagic species dwells in rivers

Description/ Biology: Upper part of the disc is dark brown with several light spots. Each of such spots is consisting of an accumulation of smaller spots. A number of separate spots are seen along the edges of the mantle. It has 2-3 rows of midtail spines (stingers) which are renewed every 6- or 12 months. Diameter of the disc is 65 cm (Fontenelle et al., 2014).

Harmful organs and harm: The presence of mid tail spines may be dangerous to humans if and when they are stepping on this fish. As it is a new species, further investigation is required on its envenomation.

Potamotrygon marquesi

Image credit: Frederik H. Mollen, Elasmobranch Research Belgium (Reproduced with permission)

Common name: It is a new species

Global distribution: Brazilian Amazon Basin

Ecology: Riverine systems

Description/ Biology: Disc of this species is rounded to slightly subcircular. Head is large and mouth is narrow Five buccal papillae are present on mouth floor. Spiracles are large and trapezoidal. Teeth are set in quincunx . Upper and lower jaws are with monognathic heterodonty. Clasper is conical and is slightly tapered distally. Tail is short and narrow. Tail is slightly dorsoventrally depressed at its base. Dorsal disc's background is in beige and grey but it is also in light to dark brown colored with open and closed bicolored rings. Ventral disc coloration is in 2-3 distinct colors. Central disc area is whitish and lateral and posterolateral margins of disc may be grey, dark grey or dark brown. Dermal denticles are found distributed over dorsal disc and tail. Tail dorsum has single row of pointed spines, from its base to caudal-sting insertion. Pointed spines are slender and are curved towards tail tip and are more developed in the midportion of the row. While total length of this species is 36 cm, its disc length is 24 cm (da Silva and Loboda, 2019)

Harmful organs and harm: Its tail and its pointed spines may be dangerous to humans if and when they are stepping on this fish. As it is a new species, further investigation is required on its envenomation.

Potamotrygon motoro

Image credit: Jim Capaldi from Springfield, USA , Wikipedia

Common name: Ocellate river stingray

Global distribution: South America : Argentina, Bolivia, Brazil, Colombia, French. Guyana, Guyana, Paraguay, Peru, Surinam and Uruguay

Ecology: Riverine systems

Description/ Biology: Based on the colour patterns, 5 groups (1-5) of this species have been identified. Among these colour patterns the 1st group which is the most known colour patter is with brown or gray dorsal disc with regular rows of tricolor ocelli (yellow, orange and black) is common in the Parana-Paraguay basin (Amazon basin). The 2nd group which is also common in Amazon basin possesses similar pattern but with bigger and varying ocelli. Ocelli of this group are bicolored (orange/red and black) with a much larger black ring. The 3rd group which is present in the Brazilian Negro-Branco river is dark colored dorsal disc with tetracolour (yellow, orange, black and beige). The 4th group which is present in Brazil and Peru has a dorsal disc with vermicular spots and broken ocelli. The 5th group which is common only in Brazil has ocelli in the form of U and is camouflaged with the disc colour (Ramos, 2017).

Harmful organs and harm: Among the freshwater stingrays, these traumatogenic species viz. *Potamotrygon motoro* and *P. scobina* are often feared by the local populations of Brazil and Peru due to the severity of the envenomation caused by the sharp, serrated four stingers located in their tail. These stingers are covered with venom-producing cells. Envenomation often occurs when these stingrays are stepped on or handled. Fishermen and bathers are mainly exposed to this danger, especially during the Amazon dry season viz. July to December. The sting of these species causes excruciating pain and marked local inflammation which is manifested as severe edema and erythema. Injuries may also be associated with skin necrosis and bacterial infections by the genus Aeromonas. Tetanus may also occur after sustaining a stingray injury (Abati et al., 2017; Brisset et al., 2006).

In this species, its robust tails with stingers may frequently cause accidents, Victims are usually stung on their lower limbs during the day while they are swimming (Haddad Jr. et al., 2013).

Kimura et al. (2014) reported that the experimental mice which had intraplantar injection of the venom of this species exhibited inflammatory reaction—including edema formation, leukocyte recruitment and release of inflammatory mediators.

Lameiras et al.(2014) reported that the stinger and dorsal extracts of this species induced local and systemic myotoxic activity

Potamotrygon orbignyi

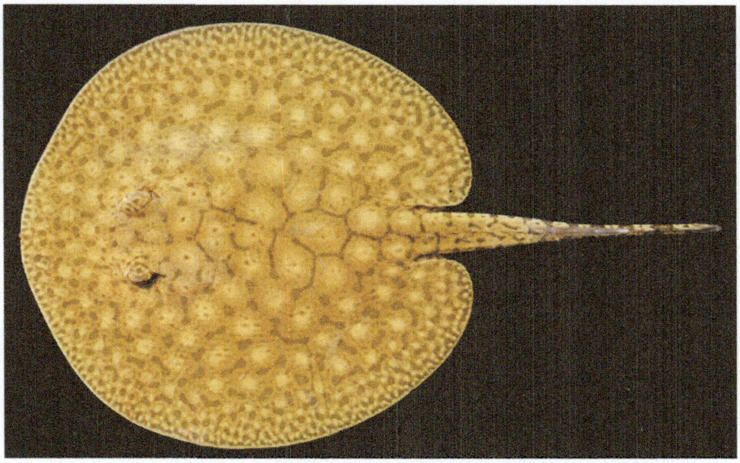

Image credit: Mark Sabaj, Perez (Reproduced with permission)

Common name: Smooth back river stingray

Global distribution: South America: Bolivia, Brazil, French Guiana, Guyana, Suriname and Venezuela; Amazon and Orinoco River basins; and Suriname, Guyana and French Guiana river systems

Ecology: Riverine systems

Description/ Biology: Coloration of its dorsal disc is varying from brown to black with beige to brown reticulated marks forming broad hexagonal or rounded patterns. Tail dorsum is with lateral vertical bands and ventrally with transversal bands. Tail is also with one or two lines of spines dorsally.

Harmful organs and harm:A vasoconstrictive nonapeptide (orpotrin) isolated from the venom of this species has been reported to reduce the diameter of mouse cremaster muscle arterioles by about 65%, with peak activity at 20 min. after injection. Further, a protein from the venom of this species exhibited minor pro-inflammatory activity (da Silva Jr. et al., 2015)

Both venom and mucus from the sting, and mucus from the skin of this species enhanced vascular permeability and promoted nociception in mice (Monteiro-dos-Santos et al., 2011)

Potamotrygon schuhmacheri

Image credit: Dreams Time (Royalty free)

Common name: No common name for this species

Global distribution: South America: Paraná-Paraguay River basin [Argentina, Brazil, Paraguay]

Ecology: This benthopelagic species dwells in riverine systems

Description/ Biology: It has a subcircular disc and its dorsal surface is with reticulate pattern of dark pigments. Maximum length of this species is 25 cm.

Harmful organs and harm: Haddad Jr. et al. (2013 reported that injuries were reported by the inhabitants and tourists who were nearby the upper course of the Paraná River

Potamotrygon scobina

Common name: Raspy river stingray, mosaic stingray

Global distribution: South America: Amazon River, Tocantins River, Pará River, and Trombetas River in Brazil.

Ecology: This benthopelagic species dwells in all freshwater bodies

Description/ Biology: It grows to a maximum length of 55 cm and its maximum weight is 15kg. Large specimens of this species had smaller and more spaced ocelli. Further a series of different color pattern has also been observed in this species

Harmful organs and harm: Both venom and mucus from the sting, and mucus from the skin of this species enhanced vascular permeability and promoted nociception in mice (Monteiro-dos-Santos et al., 2011). Vide *Potamotrygon motoro* also

Potamotrygon wallacei *(= Potamotrygon hystrix)*

Common name: Porcupine stingray

Global distribution: South America: Rio Negro drainage (Amazonas, Brazil)

Ecology: This benthopelagic species dwells in rivers

Description/ Biology: Dorsal surface of disc of this species is light brown, with black irregularly-shaped vermiculate markings which form a ?-shaped figure on mid-disc. It is the smallest species of *Potamotrygon* with only 31 cm disc width. There are 1-3 irregular rows of dorsal spines on tail. These spines are low and are extending posteriorly only to tail mid-length and not to caudal stings.

Harmful organs and harm: Its dorsal spines on tail are usually low and are in one to rarely three irregular rows. But these spines are extending posteriorly only to tail mid-length and not to caudal stings. (https://www.fishbase.in/summary/68307) The presence of few tail spines in this species may be dangerous to humans if and when they are stepping on this fish. As it is a new species, further investigation is required on its envenomation.

5.1.2. Teleost fishes

Thalassophryne amazonica **(** *= Potamobatrachus trispinosus*)

Image credit: 5snake5, Wikipedia

Subclass: Teleostei   Order: Batrachoidiformes   Family: Batrachoididae

Common name: Amazon Toadfish, Prehistoric Monster Fish

Global distribution: South America: it is confined to the Amazon River and its tributaries.

Ecology: This demersal fish is living only in freshwater habitats

Description/ Biology: It is the only species in which dorsal, caudal, and anal fin are connected with each other. It lacks a prominent lateral line on head and body where a few pores are found scattered. It grows to a maximum length of 10 cm only.

Harmful organs and harm: Both opercular and dorsal glands of this species are venomous (https://www.fishbase.se/summary/53559). Its well developed venomous apparatus consisting of venomous glands which are located on the base of two hollow prickles (derived from dorsal fin spines) on the back and two additional prickles present on the operculum. The sting of this species may cause strong pain, inflammation and sometimes necrosis, but no fatalities or major accidents have been reported on this fish.

Hydrocynus goliath **(** = *Hydrocynus vittatus ; Hydrocynus vittiger)*

Image credit: Cedricguppy - Loury Cédric , Wikipedia

Order: Characiformes Family: Alestidae

Common name: Goliath tigerfish, giant tigerfish, African Tiger Fish

Global distribution: Congo River Basin; and Lake Tanganyika.

Ecology: This pelagic species dwells in all freshwater bodies

Description/ Biology: This large-toothed and more massive jawed fish grows to a maximum length of 1.5 m and a weight of 50 kg. Its characterized teeth fit into distinct grooves along the jaws. Each of its teeth may grow up to 2.5 cm. Its dorsal fin which is in line or slightly in front of pelvic fins, very short gill rakers, and black adipose fin are the other salient features of this species.

Harmful organs and harm: It is the only African freshwater fish known to attack humans, and several incidents have been reported from the Congo River (https://www.fws.gov/fisheries/ans/erss/uncertainrisk/ERSS-Hydrocynus-goliath-FINAL-September2018.pdf).

Pygocentrus nattereri

Image credit: Carlos Henrique dos A. dos Santos (Reproduced with permission)

Order: Characiformes Family: Serrasalmidae

Common name: Amazonian red-belly piranha

Global distribution: It is native to South America, and is found in the Amazon, Paraguay, Paraná and Essequibo basins and coastal rivers of northeastern Brazil

Ecology: This pelagic species thrives in all freshwater habitats such as rivers, wetlands, floodplains and reservoirs

Description/ Biology: Adult fishes of this species have a reddish tinge in the belly and they can reach up to 3.9 kg in weight and 50 cm in length. Rest of the body is often grey with silver-flecked scales. Occasionally, blackish spots appear behind the gills and the anal fin is normally black at the base. Pectoral and pelvic fins vary from red to orange in color. Females are distinguished from males by the slightly deeper red color in their bellies.

Harmful organs and harm: It is said to be a bloodthirsty and mindless killer but according to reports , it is a shy and retiring fish. These fishes have been reported to attack humans only when they are threatened or in extreme hunger.

Pygocentrus piraya

Image credit: Wikipedia

Common name: Piraya piranha, San Francisco piranha, or man-eating piranha

Global distribution: South America : São Francisco River basin in Brazil.

Ecology: This pelagic species thrives in all freshwater habitats such as rivers, wetlands, floodplains and reservoirs

Description/ Biology: This species is one of the largest piranhas, reaching a maximum length of 50 cm. It has orange to yellow belly coloration, silver eyes, and rayed fibrous adipose fin. As most other piranhas, it is laterally compressed and fairly circular inshape , and it bears very sharp teeth. Lower jaw is thick, strong, and protruding.

Harmful organs and harm: It is a very aggressive.species This behaviour combined with its large size make it a danger to humans (Wikipedia). It is also traumatogenic species (https://www.fishbase.in/summary/Pygocentrus-piraya.html). It is also quite capable of inflicting serious wounds on a victim while handling (https://www.practicalfishkeeping.co.uk/features/piraya-piranha-pygocentrus-piraya/)

Serrasalmus rhombeus

Image credit: Wikipedia

Order: Characiformes Family: Serrasalmidae

Common name: Redeye piranha, black piranha, white piranha, spotted piranha or yellow piranha,

Global distribution: Northern South America: Amazon and Orinoco River basins; Guiana Shield rivers, and Brazilian coastal rivers.

Ecology: This bentho-pelagic fish lives in all freshwater bodies such as drainage systems, rivers, river channels and stiller stretches with thick submerged or marginal vegetation

Description/ Biology: It is one of the larger species of piranha with a rhombus-shaped body, red eyes, powerful jaws with razor-sharp, interlocking teeth. Its body color changes from silver in juveniles to black or dark purple in large adults. It grows to a maximum length of 41.5 cm and weight of 3 kg.

Harmful organs and harm: It is a traumatogenic fish. (https://www.fishbase.de/summary/Serrasalmus-rhombeus.html) . Its powerful jaws have been reported to cut out a good piece of human flesh with a single bite. It has been implicated in several attacks on bathing humans in two river resorts on the Lower Suriname River and two Amerindian villages on the Wayombo River (Mol, https://doi.org/10.1163/9789004207653_009)

Electrophorus electricus

Image credit: Stan Shebs, Wikimedia Commons

Order: Gymnotiformes, Family: Gymnotidae

Common name: Electric eel

Global distribution: It is commonly distributed in the northeastern portions of South America. Including Guyanas and Orinoco Rivers and Amazon basin

Ecology: It inhabits mainly on the muddy bottoms of rivers and rarely in swamps.

Description/ Biology: Electric eels possess strong physical resemblance to true eels. It has a dorso-ventrally depressed skull. Body is long and snake-like and caudal, dorsal and pelvic fins are absent. Body length is up to 2.5 m. An extremely elongated anal fin is present. Body is cylindrical in shape with a slightly flattened head and large mouth. Electrical organs are seen in the posterior portion of the body. Gills are present but they are not the source of oxygen intake. These fishes are obligatory air breathers as they get almost 80 % of their oxygen through their highly vascularized mouth. A thick, slimy skin covers the entire body of the fish. Color of this species ranges from gray to brownish/black with some yellowish coloration on the anterior ventral portion of the body.

Electricity production in electric eels: Electric eel has been reported to have 6000 cells (electrocytes) which are responsible for the production of electricity. Each fish can generate a power of about 600 volts, which is said to be 5 time stronger than the electricity produced in the standard US wall socket. Further, its electric shock is strong enough to knock down a horse or a rhinoceros. A man may die if an eel hits him with several electric shocks. It is also said that a 600 cm long eel can produce enough electricity to light 12 light bulbs. Since these electric eels have poor eyesight, they produce low-level electric charge i.e up to 10 volts which helps them see their surroundings and locate prey (Eissa, file:///C:/Users/rsant/Downloads/DangerousFishesArticle-ByAlaaEdinEissa.pdf)

Harmful organs and harm: Electric eels are the strongest living bioelectricity generators. These fishes possess three electric organs viz. the main organs, Hunter's organ, and Sach's organ which are made up of modified muscle cells. Main electric organ is located on the dorsal side and Hunter's organ which parallels the main organ is located on the ventral side. These two organs generate high-voltage pulses. Sach's organ, which produces lower-voltage pulses is present in the rear quarter of the electric eel and this organ allows the electric eel to communicate and navigate waters. Sach's organ has also the electric eel's negative pole. Electric eels are very dangerous to humans because of their strong electric capabilities. In this species , the electrical organs which are derived from modified muscle tissue.are three on each side of the body. It has been reported that this fish produces high-voltage EOD (electric organ discharge) i.e 350 to 650 volts which is strong enough to severely injure even humans (//nypost.com/2019/09/10/scientists-discover-shocking-high-voltage-electric-eels/). However, de Santana (10.1038/s41467-019-11690-z.) reported that its highest EOD was only 480 volts

Electrophorus varii

Image credit: Wikimedi Commons

Common name: Electric eel

Global distribution: This species is found distributed from the lowland Amazon Basin.

Ecology: This benthopelagic fish is found in the slow-flowing lowland Amazon basin waters. It prefers turbid rivers with little oxygen and sandy or muddy bottoms.

Description/ Biology: Electric eel has an elongated, cylindrical body. Its body coloration is dark gray-brown on the back and yellow or orange on the belly. Mouth is squarish and is located at the end of the snout. Anal fin extends to the tip of the tail. It has no scales. This species has a deepened skull unlike its related species and it grows to maximum length of about 150 cm.

Harmful organs and harm: It has been reported that this fish produces high-voltage EOD (electric organ discharge) i.e 572 V . Though the electric discharge of this species seems to be dangerous , it does not carry current more than 1 Ampere thereby making it safe for humans (https://in.mashable.com/science/6676/researchers-have-found-the-worlds-most-powerful-electric-eel-in-the-amazon)

Electrophorus voltai

Image credit: Wikipedia

Common name: Electric eel

Global distribution: South America; Brazilian Shield which encompasses the south of Para and Amazonas; Rondonia and north of Mato Grosso

Ecology: This benthopelagic species is found in the Amazon River or its tributaries

Description/ Biology: This species is distinguished by its ventral ,ovoid outline of head ; 112-146 lateral-line pores; depressed skull and cleithrum which lies between vertebrae 5 and 6 ; and wide head. It attains a maximum length of 171 cm.

Harmful organs and harm: Although 250 species of fish in South America are known to generate electricity, it is the only electric eel which is very dangerous to even humans. This species has been named after Alessandro Volta, the inventor of the electric batter and it , is the strongest bioelectric generator with the capacity to produce 860 volts.. While this species is believed to be lethal to humans (https://www.timesnownews.com/the-buzz/article/newly-discovered-electric-eel-is-the-most-powerful-can-generate-860-volt-shock/488043), it is also reported that these fishes do not carry current more than 1 Ampere making it safe for humans. (https://in.mashable.com/science/6676/researchers-have-found-the-worlds-most-powerful-electric-eel-in-the-amazon). According to some reports, human deaths due to electric eels are extremely rare and their , multiple shocks can cause respiratory or heart failure, and people have been known to drown in shallow water after a stunning jolt (https://www.nationalgeographic.com/animals/fish/e/electric-eel/)

Atractosteus spatula

Image credit: Greg Hume , Wikimedia

Order: Lepisosteiformes Family: Lepisosteidae

Common name: Alligator gar, gemfish

Global distribution: It is common in the Ohio River in southwestern Ohio and the Mississippi river south to the Gulf of Mexico. It is also found in parts of Florida, Georgia, Alabama, Tennessee, Arkansas, Oklahoma, and Texas

Ecology: It is found in large lakes, rivers, and bayous. Typically it is seen in backwaters and bottomland swamps. It may also occur in brackish waters and rarely enter marine waters (https://animaldiversity.org/accounts/Atractosteus_spatula/). It is one of the most resistant fish in the world, which manages to survive situations normally deadly for other species (http://www.aquascapinglab.com/en/2018/01/24/alligator-gar-atractosteus-spatula-facts-info/)

Description/ Biology: It is the largest freshwater fish in North America and is growing to about 3 m long and 137 kg. It is grayish green to brown color on its dorsal surface and yellowish or white ventrally. It has long, slender body and its jaws are armed with many teeth. Eyes are small and a heterocercal tail is present. Its swim bladder can function as a lung. Snout is short and broad with two rows of teeth on the upper jaw. Body is protected by a thick set of ganoid scales.

Harmful organs and harm: It is one of the 10 most dangerous freshwater fishes (https://www.look4ward.co.uk/awareness/10-of-the-most-dangerous-freshwater-fish/) This species has been reported to cause problems for humans in sport fishing and consumption. Several undocumented reports of injuries to humans are available. Its eggs are also poinsonous if consumed (https://animaldiversity.org/accounts/Atractosteus_spatula/)

Noturus stigmosus

Image credit: Wikipedia

Family: Ictaluridae

Common name: Northern madtom

Global distribution: North America

Habitat: This demersal fish lives in all freshwater bodies such as large rivers and lakes. It is known to prefer strong currents with sand, gravel, or rocky substrates. This species is also somewhat tolerant of turbidity while it avoids heavily silted areas.

Description: It is a stocky fish with four broken dark saddle bands, the last of which is seen at the base of the adipose fin. Dorsal fin has a brownish or dusk bar. A pair of light spots are seen just anterior to the base of the dorsal fin. It grows to a maximum length of 13 cm.

Harmful organs and Harm: The venom glands of this species are located in association with the sharp, bony spines along the leading edge of its dorsal and pectoral fins. When a spine enters a predator, the integument surrounding the venom gland cells is torn, thereby releasing its venom into the wound of the victim. Venom of this species has been reported to exhibit neurotoxic and hemolytic properties and can produce a variety of other symptoms such as severe pain and swelling at the site of envenomation in humans (Wright,2009; https://www.futurity.org/just-your-garden-variety-poisonous-catfish/)

Pristis pristis

Image credit: SFU University Communications , Fishes of Australia (CC)

Order: Pristiformes   Family: Pristidae

Common name: Freshwater sawfish, largetooth sawfish and southern sawfish

Global distribution: From the Caribbean to Central and South America; and Africa

Habitat: This species occurs both in saline coastal waters and freshwater lakes. It has been recorded from the Lake Nicaragua at a depth of more than 120 m.

Physical Description: Head of this species is ventrally flattened with the mouth and gills located underneath. Eyes are located dorsally. It has a distinct lower lobe on its caudal fin. The characteristic feature of this species is its long flat rostral "saw" which is studded with powerful teeth along its margins.it attains a maximum length of about 8 m.

Harmful organs and harm: Captured sawfish need to be handled with care as its extremely dangerous saw may cause serious injury when it is frightened or captured (https://roaring.earth/these-are-the-deadliest-freshwater-fish/)

Catfishes

There are more than 1600 species of catfishes which are seen both in marine and freshwater environments. Among them at least 1,250 species are believed to be venomous.Though most catfish use their venom for defense , some species of North America have been reported to inflict a sting in humans. Further, a few catfish species can even kill men.

Order: Siluriformes Family: Ariidae

Cathorops agassizii

Image credit: Rotundo, Matheus M., (Reproduced with permission)

Common name (s): Not designated

Global distribution:. Tropical; South America: Guyana and Brazil

Ecology: It is a freshwater, benthopelagic species.

Description: It reaches a maximum length of 32 cm. No other information is available.

Venomous organs and Envenomation: It is a traumatogenic species (https://www.fishbase.se/summary/Cathorops-agassizii.html) . Junqueira et al. (2006) reported that the stings are often contaminated by Gram negative bacterial species viz. *Klebsiella pneumoniae, Enterobacter* sp , *Escherichia coli* , *Serratia marcescens, Serratia* sp. and *Proteus mirabilis*. This sort of contamination caused environmental public health conditions, as well as, accidents with these fish stings developed significative acute secondary infections in humans.

Cathorops aguadulce (= *Arius aguadulce*)

Image credit: Wikipedia

Common name (s): Estuarine sea catfish

Global distribution: Central America: Atlantic draining rivers from Panuco River basin in Mexico to Izabal Lake in Guatemala

Ecology: It is a demersal species found in large to medium-sized rivers, lagoons and small drainages. It may also occur in brackish waters.

Salient features: Head of this species is long and depressed; and snout is long. A shield is seen on top of head. Mouth is small and its lower jaws are arched. Tail fin is deeply forked. It reaches a maximum length of about 30cm

Venomous organs and Envenomation: The potential venom delivery system which is the bony spine of this species is Y shaped. Paired anterior venom glands are present (Wright, 2012) . (https://www.fishbase.se/summary/Cathorops-aguadulce) reported that it is a traumatogenic species and the posterior margin of its pectoral-fin spine is with long and conspicuous serrations.

Potamarius nelsoni

Image credit Alexandre P. Marceniuk & Naércio A. Menezes (Reproduced with permission)

Common name (s): Lacandon sea catfish

Global distribution:. Tropical; Central America: Mexico.

Ecology: Freshwater; It is a demersal species inhabiting only freshwater systems.

Salient features: In this species males carry the eggs laid by the females in their mouth. It has a maximum length of 39 cm.

Venomous organs: It is a traumatogenic species (https://www.fishbase.in/summary/51770). Its potential venom delivery system which is the bony spine of this species is Y shaped. Anterior venom glandular cells are visible (Wright, 2012).

Centromochlus heckelii

Image credit: Mark Sabaj (Reproduced with Permission)

Common name (s): Driftwood Catfish

Global distribution: Tropical; South America: Amazon and Orinoco River basins.

Ecology: It occurs in the main course of the rivers and adjacent lagoons.

Salient features : It has large eyes and its large pectoral fins are also large and are always spread. Its dorsal and pectoral fins are provided with long and sharp serrate spines. It is known to attain a maximum length of 15 cm.

Venomous organs and Envenomation: This spiny catfish have 3 well developed and sharp spines with serrated edges, 1 in the dorsal fin and 1 in each pectoral fin, which it spreads out and locks readily when frightened . This defensive behavior may result in this catfish becoming attached by its spines to several places on a bather's body (Sazima, et al., 2005). The most commonly injured body places due to the sting of this fish were the upper limbs (mainly hands), followed by head and neck and lower limbs. Sazima, et al. (2005) reported that the serrate spine edges rendered the extraction of the attached fish difficult and painful. Further, as the skin and subcutaneous tissue was pulled during the process, the pain subsided in about half an hour. The puncture wound was found accompanied by local reddening and a slight swelling without further complications.

Clarias batrachus

Image credit: Francis Day , Wikimedia Commons

Common name (s): Walking catfish, Philippine catfish

Global distribution: Asia: Java, Indonesia.

Ecology: This freshwater, brackish, demersal and potamodromous species inhabits lowland streams, swamps, ponds, ditches, rice paddies, and pools

Salient features: It has an elongated body which is compressed posteriorly. Upper jaw is a little projecting. It is often covered laterally in small white spots Body coloration is gray or grayish brown. It attains a maximum length of 50 cm and weight of 1.2 kg.

Venomology: Spine of pectoral fins rough on its outer edge and serrated on its inner edge (https://www.fishbase.se/summary/3054)

Male *Clarias batrachus* has been reported to possess venomous pectoral and dorsal spines and its venom is mild compared to some African species, which can admit the victim in hospital (https://www.myaquariumclub.com/are-clarias-catfish-poisonousi-have-read-online-that-there-are-some-cases-...-507056.html).

Venkaiah and Lakshmipathi (2000) reported on the biochemical composition of epidermal secretions and poisonous spine extracts of this species. The protein content of epidermal secretions was found to be higher than that of the poisonous spine. the difference is about 8 folds. Trichloroacetic acid (TCA) soluble peptides, ninhydrin positive substances show significant variation between the two tissues of this species. Further, the electrophoretic patterns exhibited more number of protein bands in the extracts of poisonous spine than those of epidermal secretions in this species. It is suggested that the concentration of proteinaceous material secreting cells around the pectoral spine has led to the evolution of venomous gland in catfish. Further, rupture of this epithelium during envenomation is the source of the venom injected into the victim's body.

Clarias brachysoma

Image credit: Ton Nientied - commons.wikimedia

Common name (s): Walking catfish , Teysmann's spotted catfish, magura

Global distribution: Asia; and it is endemic to Sri Lanka.

Ecology: This freshwater and demersal species inhabits rivers, tanks , pools with muddy water

Salient features: In this species, body is laterally compressed and elongated (ribbon like) and is with out scales. Head is more depressed. snout is rather broad and is with small eyes. Mouth is terminal and teeth are in villiform bands on jaws. Pectoral spine is toughened. Body colouration

Venomology: It is a venomous species (https://www.fishbase.se/summary/13394)

Clarias fuscus

Image credit: Dr.Kwang-Tsao Shao (Reproduced with permission)

Common name (s): Hong Kong catfish, whitespotted clarias, Chinese catfish,

Global distribution:. Subtropical; Asia: China, Taiwan, Philippines and Vietnam.

Ecology: This demersal fish occurs in streams, ponds, ditches and reservoirs.

Salient features : In this species, body coloration is yellowish brown or dark gray. Its rounded tail and long dorsal and anal fins- are the characteristic features of this species. It attains a maximum length of 40 cm.

Venomous organs: : It is a venomous species. (https://www.fishbase.se/summary/Clarias-fuscus)

Family: Doradidae

Acanthodoras spinosissimus

Image credit: Clinton & Charles Robertson , Wikipedia

Common name (s): Chocolate talking catfish, talking catfish

Global distribution: South America: Amazon and Essequibo River basins.

Ecology: This freshwater, demersal species is inhabiting slow-moving and standing waters including tributaries, backwaters, freshwater swamps and coastal mangroves.

Salient features: Head of this species is more wide than long. Body is completely assembled with spines which are fairly deep covering more than half of the body and are almost in contact in the dorsal area of this fish. It attains a maximum length of 15 cm.

Venomology: It is a venomous species and is said to produce toxic fluids (https://www.fishbase.de/summary/acanthodoras-spinosissimus.html)

Sodre et al. (2021) reported that *Acanthodoras* is the only genus of catfish secreting a distinct and abundant milky-looking toxic substance through an axillary pore present just below the base of the posterior cleithral process. Histological analysis of preserved specimens of this species showed the presence of a saccular axillary gland with large, binuclear secretory cells. These cells which are located near the lumen appear to lose nuclei become filled with secretory products, possibly with proteinaceous substances.

Centrochir crocodili

Image credit: Flickr, Cornell University Museum of Vertebrates (Reproduced with permission)

Common name (s): Not reported

Global distribution: South America

Ecology: It is a freshwater and demersal species.

Salient features : It is known to attain a maximum length of 23 cm and weight of about 250g.

Venomous organs: It is a venomous species and it has caused human envenomations especially in South America (http://www.wikitox.org/doku.php?id=wikitox:penetrating_marine_envenoming).

Pterodoras granulosus

Image credit: Johann Natterer ,Wikipedia

Common name (s): Granulated catfish

Global distribution: Subtropical; South America: Guyana and Suriname.

Ecology: This demersal species which occurs in small groups lives in riverine systems. It is a migratory species travelling between downstream and upstream during March-July and October-Jnauary.

Salient features: It has a scaleless body but its skin is thick and tough. Body coloration of this species may vary depending upon the location albeit it usually has a muddy-brown colour with some darker spots over the body and fins. Eyes are very small. Caudal fin is deeply forked. This species grows to maximum length of 70 cm and attains a weight of 6.5 kg.

Venomous organs: It is a venomous species and it has caused human enveno-mations in South America (http://www.wikitox.org/doku.php?id=wikitox:penetrating_marine_envenoming).

Brachyrhamdia marthae

Image credit: Worldfish Wiki

Common name (s): Three-barbeled catfishes

Global distribution: Neotropical; South America: Coratijn, Essequibo, and Negro River basins.

Ecology: These are demersal species

Salient features: These are small siżed fishes with remarkable color patterns.. It attains a maximum length of 4.5 cm.

Venomous organs: In this species, the venom glands are associated with the pectoral fin. These venom glands include both surface glands of the ossified pectoral-fin spine and axillary glands the secretions of which are toxic (Carvalho, et al., 2021). The cells of the pectoral- fin spine glands are found concentrated at the posterior margin of spine, near the posterior serrae.

Family: Heteropneustidae

Heteropneustes fossilis

Image credit: Francis Day and Suzini - Wikipedia

Common name (s): Stinging catfish

Global distribution:. Asia: Pakistan and Sri Lanka to Myanmar.

Ecology: This demersal species is commonly found in muddy rivers, flooded lakes,, ditches, swamps, and marshes. It can tolerate slightly brackish water. Body of the species is uniform grey-brown to olive-brown. Two narrow, pale to yellowish, longitudinal bands are present and there are numerous black spots on its flanks. It grows to a length of about 30 cm.

Venomous organs: This species is considered very dangerous to humans, as its pectoral fin spine is very strongly serrated and connected to a poisonous gland (Thalgaspitiya, et al., 2017).

Heteropneustes microps

Image credit: Wikipedia

Common name (s): Small Asian Stinging catfish

Global distribution: Tropical; Asia: Sri Lnaka, India

Ecology: Adults occur mainly in irrigation ditches and swamps .This demersal, freshwater species may enter brackish water also

Salient features: Body depth of this species equals to head length. Its terminal mouth has well developed lips Pectoral spine is separated from rays and is serrated along its inner edge. It attains a maximum length of about 20 cm.

Venomous organs and Envenomation: This is able to deliver a stinging protein (known as poison) released from the venom glands located around its pectoral spine. In humans, this catfish venoms, which are reported to be neurotoxic and hemolytic (i.e. destruction of red blood cells), may produce local numbness, inflammation and severe painful physical conditions of the human limb when the latter is under attack (https://www.fws.gov/fisheries/ans/erss/uncertainrisk/ERSS_Heteropneustes_microps_final_January_2018.pdf)

Ameiurus catus

Image credit: Ellen Edmonson and Hugh Chrisp , Wikipedia

Common name (s): White catfish

Global distribution: North America: Rivers of the Atlantic coastal states of USA (Florida to New York).

Ecology: This freshwater, brackish, demersal and potamodromous species is largely found in sluggish, mud-bottomed pools, open channels, and backwaters.

Salient features: This species is normally gray to blue-black above and white to light yellow below. It is largely distinguished by its dusky or black adipose fin, shorter anal fin base, and lower degree of forking in the caudal fin. Head is with eight barbels, two nasal, two maxillary and four chin. It attains a maximum length of 95 cm .

Venomology: It has a venomous spine on the anterior edge of its dorsal and pectoral fins. A maximum of 15 fairly large saw-like teeth is seen on the rear edge of its s pectoral fin spine. The venom of this species was found to contain two to eight fractions that showed lethal activity in mice. Further, two putative toxic peptides prepared from the fin spine extracts of this species produced a significant toxic effect in injected largemouth bass. The number of lethal fractions in the venom of this species is much more consistent with what is known from other species. (Wright, 2015)

Noturus baileyi

Common name (s): Smoky madtom

Global distribution: Temperate; North America: Tennessee, USA.

Ecology: This demersal species occurs in rocky riffles, runs, and flowing pools of creeks. This species relies largely on clean water and undisturbed rocks to complete its life cycle

Salient features: It is olive brown on dorsally and belly is white to yellow. Four saddles are found to line the back of this fish. it is a small sized fish with a total length of only 7 cm.

Venomous organs: It has venomous spines that can deliver a painful sting to humans. Its sting is generally considered equivalent to a bee sting (https://www.wbtv.com/2019/10/16/venomous-fish-with-painful-sting-was-presumed-extinct-smoky-mountains-its-back).

Noturus exilis

Image credit: Wikipedia

Common name (s): Slender madtom

Global distribution: Temperate; North America: central Kentucky to northern Alabama; from southern Wisconsin and southern Minnesota to Ozark and Ouachita Highlands of Arkansas, Kansas and Oklahoma in USA.

Ecology: This demersal fish inhabits small rivers., rocky riffles, and flowing pools of clear creeks.

Salient features: This long and slender catfish has a scaleless body and it has eight distinct barbels around its mouth. Body coloration is continuous grey to dusky yellow dorsally with the exception of two light yellow spots viz. a large spot on its nape, and a smaller one immediately behind dorsal fin. Belly is light yellow to whitish. It feeds on aquatic insects and small crustaceans such as isopods, amphipods and copepods. It has a maximum length of 15 cm.

Venomous organs and Envenomation: The serrae present on its pectoral spines may give most painful stings. These serrae may inflict wounds that can cause extreme swelling for several days (https://animaldiversity.org/accounts/Noturus_exilis/). Extracts of the stings of this species have also been reported to be lethal to Gambusia fish (Birkhead,1972).

Noturus gyrinus

Image credit: Ellen Edmonson and Hugh Chrisp , Wikipedia

Common name (s): Tadpole madtom

Global distribution: Temperate; North America: from New Hampshire to Texas and Gulf of Mexico, USA; Quebec to Saskatchewan in Canada.

Ecology: This demersal species inhabits lakes, rock, mud or detritus-bottomed pools and backwaters of creeks and small to large rivers

Salient features: Body of this species is heavy, round, and potbellied anteriorly and is strongly compressed posteriorly . Its terminal mouth is with numerous small and sharp cardiform teeth. It has a dark brown back with a lighter brown color on their sides and a yellow or white belly. It reaches a maximum length of 13 cm.

Venomous organs: It is one of the more venomous catfishes with toxic (neurotoxic and haemotoxic) sheath covering dorsal and pectoral spines. Threatened fish will arch back and erect its sharp spines. Effects of stings of these spines on humans are comparable to that of bee stings (http://txstate.fishesoftexas.org/noturus%20gyrinus.htm).

Family: Malapteruridae

Malapterurus beninensis

Common name (s) : Electric catfish

Global distribution: Africa

Ecology : It is a freshwater, benthopelagic species. It mostly occurs in lowland habitats such as marshes, rivers, and lakes

Salient features: Body of this species is cylindrical with a large head and a thick caudal peduncle. Eyes are relatively small. Dorsum and flank nearly always spotted. It attains a maximum length of 25 cm. Electric catfish have been reported to achieve immunity by their highly resistive tissues which shield the whole animal or its muscles, heart and nervous system. Further, it is said that some tissues might have also evolved some intrinsic tolerance (Welzel and Schus,ter, 2021)

Harmful effects: The most notable feature of this genus is its strong electrogenic ability. The electric organ has been reported to evolve from the pectoral muscle and it surrounds the body over most of the length of the fish. This electric organ is capable of discharging up to 350 V in a 500 mm fish .(https://animaldiversity.org/accounts/Malapterurus_electricus/)

The electric shocks given from specimens of even only 4-5 cm long are have been reported to be painful. The electric shocks of this species may reach 350-450 volts ((https://www.aquariumglaser.de/en/fish-archives/malapterurus-beninensis-electric-catfish/). It is advisable therefore not to touch this fish..

Malapterurus electricus (=*Torpedo electricus*)

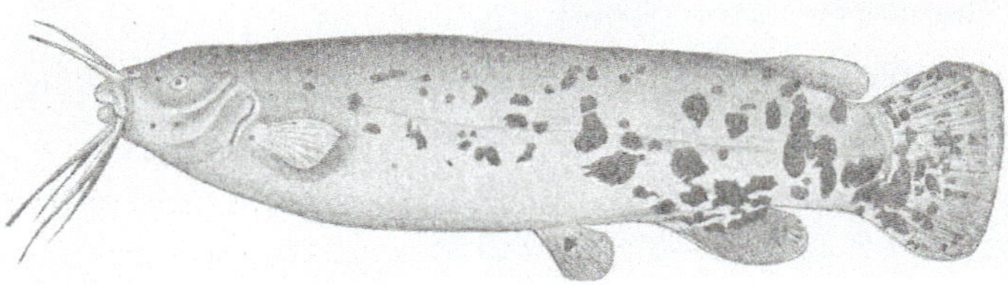

David Starr Jordan, Wikipedia

Common name (s) : Electric catfish

Global distribution: Africa

Ecology : It is a freshwater, benthopelagic and potamodromous species. It is generally found among rocks or roots in turbid , sluggish or standing water

Salient features: Body of this species is elongate and cylindrical with a slightly depressed. Eyes are small and snout is rounded or blunt. Mouth is relatively wide. Both adults and young are marked with large spots and blotches. It grows to a maximum length of about 120 cm and weight of 20 kg.

Harmful effects: Its electric organ which is capable of discharging 300-400 V is derived from pectoral muscle and it surrounds almost the entire body. Its EOD (electric organ discharge) duration is 1.5 ms (https://www.fishbase.de/summary/2087) . Though this species is giving an unpleasant surprise to the occasional careless fisherman who handles it, its electric shock is not known to be fatal to humans (https://animaldiversity.org/accounts/Malapterurus_electricus/)

Malapterurus microstoma

Yuuki 315, commons.wikimedia.

Common name (s) : Small mouth electric catfish, dwarf electric catfish

Global distribution: It is native to the Central African Republic, the Democratic Republic of the Congo and the Republic of the Congo

Ecology : it is a freshwater, benthopelagic species

Salient features: Body of this species is finely spotted and its lateral line is complete. Mouth is relatively narrow and snout is relatively long and sharp. Its horizontally based pectoral fins are found low on the body. It grows to a maximum length of 60 cm.

Harmful effects: The electric organ of the adult fish has been reported to generate 300-400 volts which is believed to be enough to stun an adult human (https://www.scotcat.com/factsheets/malapterurus_microstoma.htm)

Malapterurus minjiriya

Common name (s) : Electric catfish

Global distribution: It is native to Burkina Faso, Ethiopia, Ghana, Mali, Nigeria and Togo.

Ecology : It is a freshwater, benthopelagic species

Salient features: Body of this species is with moderately to very large black spots which are irregularly scattered. Its oral tooth bands are much larger compared to M.electricus. It grows to a maximum length of about 1 m.

Harmful effects: In the generation of electric shock, it largely resembles M. microstoma

Family: Mochokidae

Synodontis njassae (= *Synodontis zambesensis*)

Image credit: Paweł Cie la Staszek_Szybki_Jest, Wikipedia

Common name (s): Njassae Synodontis Catfish ; Malawi squeaker; upside-down catfish

Global distribution:. It is endemic to Lake Malawi, East Africa

Ecology: It occurs in the inflowing streams and is known to be abundant on rocky shores.

Salient features: It has characteristic small, medium, and large spots on its body. Head is large and bony. It attains a maximum length of 30 cm.

Venomous organs: The potential venom delivery system which is the bony spine of this species is Y shaped. Paired anterior venom glands are seen (Wright, 2012). It has a strong, sharp, serrated spine at the front of the dorsal fin and in the pectoral fins. These spines are found locked at right angles to the body. The mucus covering these spines hs been reported to be toxic (https://www.scotcat.com/articles/article119.html).

Family: Pangasiidae

Pangasianodon hypophthalmus (= *Pangasius sutchi*)

Image credit: Wikimedia Commons

Common name (s): Striped catfish; iridescent shark; shark catfish

Global distribution:. Tropical; Asia: Mekong, Chao Phraya, and Maeklong basins.

Ecology: This benthopelagic species inhabits large rivers it may also thrive in rice fields.

Salient features: Body of this species is elongated and its head is relatively small. Mouth is broad and is in terminal in position. Eyes are large. Fins are dark grey or black. Body of adult fishes is uniformly grey in colour. It has a maximum length of 130 cm.

Venomous organs and Envenomation: The potential venom delivery system which is the bony spine of this species is Y shaped. Paired, hemispherical venom glands are seen in the spine (Wright, 2012). Human envenomation due to this species has been reported as a case. A 51-year-old Malay man was brought to the emergency department because of his left hand pain caused by a puncture wound from the spine of this species (Ahmad et al., 2013)

Family: Pimelodidae

Brachyplatystoma capapretum

Common name: Piraiba catfish, dark caped goliath catfish

Global distribution: Tropical; South America: Brazil and Peru

Ecology: This benthopelagic species commonly inhabits muddy waters and deeper, flowing channels at a depth range, 5 - 10 m.

Salient features: Body coloration of this species is black to dark gray or brown dorsally and it is plain white to dusky cream ventrally. Caudal fin is moderately forked. It attains a maximum length of about 100 cm and weight of > 30 kg.

Venomous organs: The shaft portion of the pectoral spines and the dorsal and anterior articular processes of the spine base are harmful. Further its hunter type of teeth are also dangerous to humans. Owing to these features, this species is also known as man-eater (https://whatdewhat.com/dangerous-freshwater-fish-can-kill/)

Pimelodus maculatus

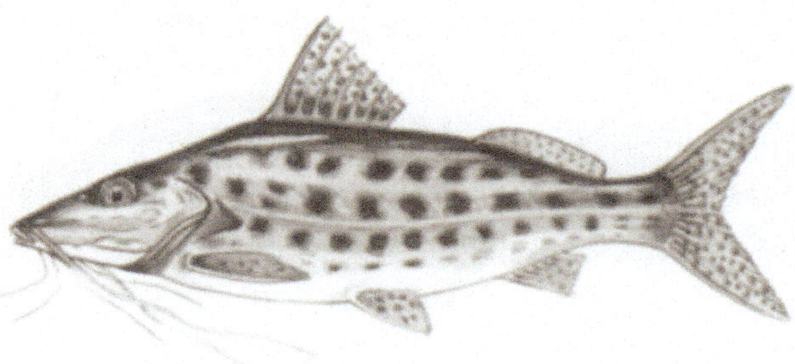

Image credit: Wickimedia

Common name: Spotted pim

Global distribution: Subtropical; South America: Paraná and São Francisco River basins.

Ecology: It is a benthopelagic, potamodromous species occurring mainly in ponds, pools and small lakes. It is also seen in shallow, flowing waters over sandy or muddy substrates

Salient features: It is identified by its large blotch patterning which covers the whole body of this fish. It attains a maximum length of about 50 cm and weight of 2.4 kg.

Venomous organs and Envenomation: This species has been reported to be the most common cause of injuries in all rivers and lakes of Brazil (de Aquino et al., 2016). The potential venom delivery system which is the bony spine of this species is Y shaped. Paired anterior and posterior venom glands are seen in spine (Wright, 2012).

Sarmiento et al.(2015) reported that this species causes frequent accidents mainly amongst fishermen. Further, they estimated that the crude venom of one fish contained about 100 μg of protein which is the quantity involved in the human envenomation. According to them, its crude venom induced significant nociceptive and edematogenic effects and caused vascular permeability alterations at doses from 30 to 100 μg/animal.Furthermore, this crude venom also induced significant changes in the levels of creatine kinase (CK) and its isoenzyme creatine kinase myocardial band (CK-MB) in mice .

Pseudoplatystoma fasciatum

Image credit: Wikimedia Commons

Common name (s) : Tiger shovelnose catfish, barred shovelnose or barred catfish

Global distribution: It is native to the Suriname, Corantijn and Essequibo

Ecology : It is seen from large rivers and lakes to areas of flooded forest. It is often found in more sluggish habitats

Salient features: It has an elongated body and a long snout. Mouth is large and looks like a duck-bill. Eyes are relatively small. Its body coloration ranges between silver and brown and belly is usually white to silver. It grows to a maximum length of 90 cm and attains a weight of about 70 kg.

Venomous organs: The crude venom of this species caused strong nociceptive and edema (swelling) effect in mice at dose de 30 μg of protein (Sarmiento et al., 2015). de Aquino et al. (2016) reported that the species of this genus caused injuries in the professional fishermen in the Corumbá and Miranda towns in the Pantanal region.

Family: Plotosidae

Neosilurus ater

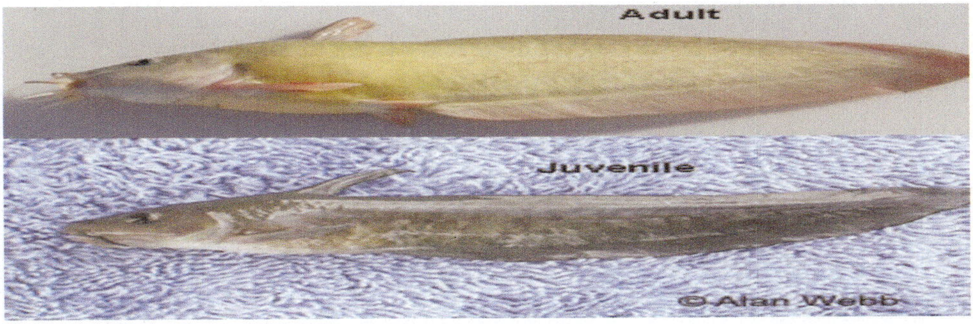

Image credit: Alan Webb, James Cook University (Reproduced with permission)

Common name (s): Narrowfront tandan

Global distribution: Tropical; Asia and Oceania: northern Australia and central-southern New Guinea.

Ecology: This solitary, demersal species is found in rivers and smaller streams with strong currents. It also lives in quiet or week flowing water of pools and side channels.

Salient features: Body coloration of this species is gray to blackish and belly is pale. It has a maximum length of about 50 cm and weight of 2 kg.

Venomous organs: It is a venomous catfish (https://www.fishbase.se/summary/Neosilurus-ater). Its dorsal and pectoral fins which are armed with serrated venomous spines may inflict a very painful sting.(https://fishesofaustralia.net.au/home/species/2762)

Neosilurus brevidorsalis

Image credit: , Allan James (Reproduced with permission)

Common name (s): Short finned tandan, short-finned catfish

Global distribution: Asia and Oceania: northern Australia and central-southern New Guinea.

Ecology: This demersal species inhabits lakes, swamps, rivers and backwaters. It also lives in flooded lagoons and billabongs.

Salient features: Body coloration is dark brown to grey dorsally and belly is white. It feeds on insects, prawns, mollusks and small crayfish and attains a maximum length of 20 cm.

Venomous organs: It is a venomous species doing harm to intruders (https://www.fishbase.in/summary/12095)

Plotosus papuensis

Common name (s): Papuan eel-catfish

Global distribution: Tropical; Asia and Oceania: New Guinea.

Ecology: This species is found in turbid rivers, backwaters and swampy lagoons

Salient features: This species is characterised by its elongated body which is compressed and tapers towards the tail end. Among its two dorsal fins, first one is bearing strong serrated complete lateral line is present. It attains a maximum length of about 100cm.

Venomous organs: The first dorsal and pectoral fins which are associated with production of venom may harm the people (Usman et al., 2013; https://www.fishbase.de/summary/plotosus-papuensis.html)

Porochilus argenteus

Image credit: Museums Victoria, Fishes of Australia, CC

Common name (s): Silver tandan

Global distribution: Tropical; Oceania: endemic to Australia.

Ecology: This demersal species largely found in waterholes in the larger rivers.; and slow-flowing streams over sand or rock bottoms.

Salient features : It is a slender bodied fish with a concave head-snout profile. Eyes are relatively low and are closer to the mouth. Its steeper head is up to the insertion of the dorsal fin. Caudal fin is fairly pointed. It has a maximum length of 35 cm.

Venomous organs: It is a venomous species (https://www.fishbase.in/summary/Porochilus-argenteus.html)

Tandanus bostocki

Image credit: Museums Victoria, Fishes of Australia (CC)

Common name (s): Freshwater cobbler

Global distribution: It is common between Moore River(Perth, Australia) to the Blackwood River in the south-west corner and as far east as Frankland River (Western Australia)

Ecology: It prefers deeper waters of larger rivers, creeks, dams and lakes

Salient features: It is easily identified by its large, eel-like tail i.e its second dorsal fin is found fused with the caudal and anal fins. Its first dorsal fin and both pectoral fins have a single venomous spine. It attains a maximum length of 90 cm.

Venomology: The venom of this species has been reported to cause extreme pain. Further, localised edema (swelling) may also result due to a hemolytic effect (destruction of red blood cells and release of fluid into bloodstream; causes increased blood flow in the area of the injury) or due to vasoconstriction (reduction in size of veins). which elicits pain due to pressure from restricted flow). The secondary effects from this catfish spines are also due to bacterial and fungal infections which may be introduced through the puncture wound (https://rivers.dwer.wa.gov.au/species/tandanus-bostocki/#:~:text=The%20mouth%20is%20surrounded%20by,venom%20gland%20cells%20is%20torn).

Tandanus tandanus

Image credit: Wikimedia

Common name (s): Freshwater catfish , eel-tailed catfish, tandan catfish

Global distribution: Temperate; Oceania: endemic to Australia.

Ecology: This demersal and potamodromous species inhabits ponds, lakes and slow moving streams with vegetation.

Salient features: Head of this species is large. Its down-turned mouth is surrounded by four pairs of barbels. First dorsal fin is high, with a strong serrated spine. Dorsum may be olive-green to brown, black or purplish; and white ventrally. It attains a maximum length of 90 cm and weight of 7 kg.

Venomous organs: The first dorsal and pectoral fins are preceded by a sharp serrated spine which may be harmful to intruders (https://www.fishbase.se/summary/Tandanus-tandanus)

Family: Siluridae

Silurus glanis

Image credit: Bernard DUPONT from FRANCE , Wikipedia

Common name: Wels catfish or sheatfish

Global distribution: It is native to central, southern, and eastern Europe; Asia.

Habitat: This benthopelagic fish lives in large, warm lakes; deep, slow-flowing rivers ; and in brackish water

Description: It has a broad, flat head and wide mouth which has several lines of numerous small teeth; two long barbels on the upper jaw; and four shorter barbels on the lower jaw. It has a long anal fin which extends to the caudal fin, and its small sharp dorsal fin is seen far forward. Pectoral fins are sharp and its skin is very slimy. Skin colour varies in relation to its living environment. While clear water gives the fish a black color, muddy water will produce green-brown individuals. Belly is pale yellow to white in color. It grows to a maximum length of 5m and weight of 300kg.

Harmful organs and Harm: This fish has been reported to eat humans when it attacks, though attacks are relatively rare (https://insmoothwaters.com/most-dangerous-freshwater-fish/#:~:text=The%20goonch%20fish%20is%20also,as%20a%20predator%20to%20humans)

Family: Sisoridae

Bagarius yarrelli

Image credit: Henry Sullivan Thomas , Wikipedia

Common name: Giant devil catfish, giant painted catfish or goonch

Global distribution: India, Nepal and other East Asian and Southeast Asian countries.

Habitat: It is commonly seen in large rivers with fast current; and in deeper pools near faster current. It hides usually amongst large rocks and boulders. The muddy color of this species helps to camouflage it well into its environment.

Description: Body of this species is elongated with a depressed head. Mouth is inferior and crescentic. Dorsal fin is inserted nearer to the adipose base than snout tip. Four pairs of barbels are seen.Eyes are circular. Caudal fin is deeply forked and its upper lobe is produced into a long filament. Skins on sides and above are rough. Body coloration is greenish or olivaceous to pronounced tan or brown. Dark pigmented bands are seen on body; first one from the near dorsal base to pelvic, second from adipose base to anal fin; and last one at caudal peduncle. Caudal is yellowish gray. Pelvic and pectoral fins are with black spots. Anal fin is with a broad black-colored band. It grows to a maximum length of 2 m and weight of 90 kg.

Harmful organs and Harm: It is a predator to humans. It has been reported that this species is responsible for a series of fatal attacks on humans in Kali river between India and Nepal between 1998 and 2007. It has developed a taste for human flesh due to its corpse feeding in this river. Swimmers are warned not to bathe in Kali River (https://roaring.earth/these-are-the-deadliest-freshwater-fish/)

Family: Trichomycteridae

Vandellia cirrhosa

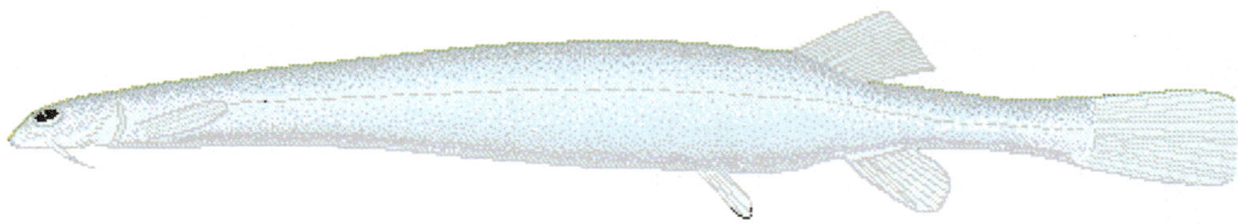

Image credit: Wikipedia

Common name: Toothpick fish, candiru

Global distribution: This demersal fish is found exclusively in the upper Amazon River and Orinoco River basins in northern South America.

Ecology: It prefers shallow, slow moving, acidic waterways with muddy or sandy bottoms. It is often found burrowed in the riverbed most of the time and is only emerging to feed or mate.

Description/ Biology: It is a small, thin catfish which has no scales. Body is transluscent and is becoming colored only after feeding. It has barbels near its mouth which is lined with minute, needle-like teeth. Body is narrow and cylindrical, with a slightly flattened head. It has short, backward facing spines on its gill covers. Its large ,black eyes are placed on top of the head. It grows to a maximum length of 17 cm.

Harmful organs and harm: Reports are available on candiru swimming up the urethra of men and women who urinate while swimming in water. It is also believed that candiru attacks are accidental and victims may die once the entered candiru is inside the urethra (https://animaldiversity.org/accounts/Vandellia_cirrhosa/)

It is also reported that candiru makes dangerous attacks on humans . It is attracted to blood and urine of humans as this species extracts its main source of food from the urine which is urea. Further this fish easily passes through anus, vagina or penis of men in water and starts sucking blood causing extreme pain to the victim (https://www.researchgate.net/profile/Alaa_Eissa2/publication/303005861)

Order: Tetraodontiformes Family: Tetraodontidae

Chelonodon patoca (= Tetraodon patoca)

Image credit: Wikimedia Commmons

Common name: Milkspotted puffer, Gangetic pufferfish

Global distribution: East Africa to the Admiralty Islands, New Britain and Trobiand Islands, north to China, south to northern Australia

Ecology: It lives in all aquatic habitats such as freshwater, brackish water and marine

Description/ Biology: Body of this species is covered with prickles. It grows to a maximum length of about 40 cm.

Harmful organs and harm: Its consumption may lead to pufferfish poisoning due to its tetrodotoxin content. Two tetrodotoxins, namely PFT-1 and PFT-2 were isolated from the liver ofthis species. Both these fractions showed pronounced effects on total RBC, WBC, platelet and ESR. Further serum levels of SGPT, SGOT, SALP, bilirubin, creatinine and urea were also found affected by these tetrodotoxins (Hasan, https://scialert.net/fulltext/?doi=pjbs.2007.773.777)

Colomesus asellus

Image credit: Oliver,S., Wikipedia

Common name: Amazon pufferfish

Global distribution: Amazon basin in Brazil, Colombia, Peru, and Ecuador

Ecology: It occurs in the lower, middle, and upper river basins with sandbars and banks with overhanging vegetation. It prefers habitats with high oxygen contents.

Description/ Biology: It grows to a maximum length of 13 cm. Coloration of the body is green above, and white below. Body is patterned with black transverse bands across the dorsal surface. It also has a distinctive black band that rings the base of the caudal fin.

Harmful organs and harm: It is known to contain the paralytic shellfish toxins viz. saxitoxin (STX) and gonyautoxin such as the GTX 2 and GTX (Oliveira et al., 2006). The toxicity of this species to humans is not yet known

Colomesus psittacus

Image credit: Wikipedia

Common name: Banded puffer

Global distribution: Amazon River in Brazil.

Habitat: This demersal fish lives in all aquatic habitats such as marine, freshwater and brackish water

Physical Description: Body excepting snout, pectoral base and caudal peduncle is covered with prickles..Teeth are fused into plates and each jaw has two such plates. Body coloration is dark green dorsally with six transverse black bars and it is white ventrally. Fins are dusky green or dark brown. It grows to a maximum length of 29 cm.

Harmful organs and harm: It has an edible flesh and a very toxic liver, but its toxins and toxicity have not been studied (https://www.scielo.br/scielo.php?script=sci_arttext&pid=S1678-91992006000200001#co).

Dichotomyctere fluviatilis (= *Tetraodon fluviatilis*)

Image credit: Steven G. Johnson , Wikipedia

Common name: Green spotted pufferfish

Global distribution: South and Southeast Asia

Ecology: It dwells in freshwater and brackish water and mangrove habitats

Description/ Biology: It has a pear-shaped body, very thick and broad forehead and two forked nasal tentacles. Body is covered with small spines. It is usually white or grey ventrally. Its body usually has a leathery texture . This species grows to a maximum total length of 17 cm

Harmful organs and harm: This species is toxic to humans due to its tetrodotoxin content (https://www.nationalgeographic.org/media/photo-ark-green-spotted-pufferfish/)

Dichotomyctere nigroviridis (= *Tetraodon nigroviridis*)

Image credit: Starseed , Wikipedia

Common name: Spotted green pufferfish

Global distribution: Asia: Sri Lanka to Indonesia and north to China.

Habitat: This demersal fish inhabits freshwater streams, rivers and floodplains; and brackish water

Physical Description: Upper half of the body of this species has many brown spots which are outlined in green-yellow. Belly is white. Its caudal fin has 1 to 8 dark transverse bands and dorsal fin may have small dark spots near the base. Other fins are however, plain. It grows to a maximum length of 17 cm. There are two flattened and broadened lobes.

Harmful organs and harm: The skin and liver of this pufferfish exhibited high concentrations of tetrodotoxin (TTX) and these findings posed a danger to small children who consume this fish (Williams et al., 2016).

Saitanu (1992) reported on the toxicity of this species. The values of toxicity of its internal organs such as skin, gonad, intestine, liver and muscle showed a decreasing trend (3231, 1064,174,163 and 85 mouse units/g respectively)

Dichotomyctere ocellatus (= *Tetraodon steindachneri; Tetraodon biocellatus*)

Image credit: Wikipedia

Common name:Palembang puffer, Eyespot pufferfish, figure eight pufferfish

Global distribution: Asia: Indochina, Cambodia, Malaysia and Indonesia.

Ecology: This demersal species lives in small streams and drainage ditches

Description/ Biology: It has a broad forehead and protruding eyes. Dorsal and anal fins are rounded and are located opposite each other. Caudal fin is fan-shaped. Skin is leathery and is covered with small spines. Belly is white and upperparts are light to dark gray. Upperparts are covered with various green to yellow patterns, ranging from lines to circles or dots to stripes. Fins are gray. It grows to a maximum length of 10 cm.

Harmful organs and harm: Saitanu (1992) reported on the toxicity of this species. The values of toxicity of its internal organs such as skin, intestine, muscle and liver showed a decreasing trend (961, 191,69 and 43 mouse units/g respectively)

Leidon cutcutia (= *Tetraodon cutcutia*)

Image credit: Wikipedia

Common name: Ocellated pufferfish

Global distribution: It is native to southern Asia from India to the Malay Archipelago

Ecology: This demersal fish dwells in freshwater bodies such as ponds, beels, canals and rivers; and brackish water areas

Description/ Biology: It has a broad head and back and is is tapering abruptly to tail. Its mouth is little inferior with two large teeth on each jaw. Gill-openings are very reduced and are restricted in front of pectoral base. Eyes are large and are situated slightly behind middle of head. Dorsal fin is located well back and above origin of anal. All fins are rounded. Body coloration is greenish yellow above and white ventrally. A light band is seen between eyes. A large black ocellus which is surrounded by a light edge is present on anterior side to the origin of anal fin. It grows to a length of 15 cm.

Harmful organs and harm: This species has been reported to be responsible for food poisoning in Bangladesh. During 1988-96, 55 people were affected and 16 died after consumption of this species (Chowdhury and Ahasan ,2015) . The TTX producing bacteria associated with this puffer such as *Bacillus, Pseudomonas* and *Staphylococcus* have also been reported to be the causative agents for the food poisoning . Among these bacteria, *Staphylococcus xylosus* which was isolated from muscle tissue of the fish at a concentration of 72 MU/g showed high toxicity (https://shodhganga.inflibnet.ac.in/bitstream/10603/151800/14/14_chapter%206.pdf)

Marilyna pleurosticta

Image credit: Gunther (copyright expired)

Global distribution: Australia

Ecology: This demersal species inhabits estuaries, muddy coastal areas and freshwater bodies. It prefers shallow waters of 1-2 m depth and vegetative areas and mangroves

Description/ Biology: It is a pale olive-green to dark greenish-grey puffer with silvery to white coloration ventrally. There are four dark bands (either solid or of irregularly distributed round spots) across the back and sides. Bands on the side of the body may however be replaced by dark, lateral blotches. Pectoral and dorsal fins are with tinged reddish to reddish-yellow coloration and anal fin is bright orange to yellow. Caudal fin is reddish-yellow,and is becoming bright orange distally. It grows to a maximum length of 15 cm.

Harmful organs and harm: Its flesh is highly poisonous and unfit for human consumption .The toxin of this fish is concentrated mainly in the ovaries, liver and intestines. It is a favourite fish in Japan and licensed chefs are permitted to prepare dish form this fish for serving in restaurants. However, the small amount of toxin present in the flesh of the fish has produced numbing effect in the mouth of the users and it is known that about more than 50 Japanese die per year by eating this fish. (https://hollowaysbeacheec.eq.edu.au/SupportAndResources/FormsAndDocuments/Documents/fish-toadfish.pdf)

Pao fangi (= *Tetraodon fangi*)

Common name: Not reported

Global distribution: Southeast Asia: Vietnam

Ecology: This demersal fish lives in all freshwater habitats

Description/ Biology: Snout of this species is relatively short. A large spot with a red core is present under the origin of the dorsal fin. Body of this species has a yellow web pattern largely. It grows to a standard length of less than 7 cm .

Harmful organs and harm: Food poisoning due to this species has been reported from Amphur Chumpuang, Nakorn Ratchasima Province, Thailand. Highest toxicity was observed in skin (813 mouse units per g) followed by eggs (336), muscle (331), liver (209) and intestine (159) (Saitanu et al., 1991). Sato et al. (1997) reported on the presence of saxitoxin and associated food poisoning of this species in Thailand

Pao palembangensis (= *Tetraodon palembangensis*)

Image credit: The Great Mule of Eupatoria , Wikipedia

Common name: Humpback / Dragon Puffer

Global distribution: Southeast Asia: Laos, Thailand, Malaysia, and Indonesia.

Ecology: This demersal species lives in freshwater bodies

Description/ Biology: It grows to a maximum length of 21 cm. No other information is available for this species.

Harmful organs and harm: Food poisoning due to this species has been reported from Amphur Chumpuang, Nakorn Ratchasima Province, Thailand. Highest toxicity was observed in skin (907 mouse units per g) followed by eggs (332), muscle (282), liver (225) and intestine (143) (Saitanu et al., 1991).

Takifugu obscurus

Image credit: Wikipedia

Common name: Obscure pufferfish

Global distribution: It is found in rivers in China and the Korean Peninsula

Ecology: This demersal and anadromous species dwells in freshwater, brackish water and marine habitats.

Description/ Biology: Body of this species covered with prickles. A large black blotch is seen on side beneath pectoral fin and on base of dorsal fin. Snout is blunt and mouth is small. A yellowish tint is seen in the fins while the fish is alive. It grows to a maximum length of 40 cm.

Harmful organs and harm:It is a poisonous species to eat. Several organs of this species have been reported to contain tetrodotoxin and therefore it is to be prepared safely by trained persons before consumption (https://www.fishbase.in/summary/Takifugu-obscurus.html)

Takifugu ocellatus

Common name: Orange saddle fugu puffer

Global distribution: Asia: China and Viet Nam.

Ecology: It is a demersal and anadromous species dwelling in freshwater, brackish-water and marine habitats

Description/ Biology: Body of this species is covered with prickles. Dark spots are seen on either side, above. Similar spots may also be seen behind pectoral fins. It grows to a maximum length of 15 cm only.

Harmful organs and harm: The fish is highly toxic and it is deadly poisonous to eat. The poison tetrotoxin is seen in the internal organs such as liver, ovaries, testes and skin of this fish . However, the flesh of this species is a delicacy in Japan and after processing by experienced chefs , this fish is served in restaurants (https://www.fishbase.in/summary/Takifugu-ocellatus.html)

Takifugu poecilonotus

Image credit: Wikimedia Commons

Common name: Finepatterned puffer

Global distribution: Japan and the southern Korean Peninsula. Reported from Taiwan

Ecology: It is a demersal and anadromous species dwelling in freshwater, brackish water and marine habitats

Description/ Biology: Body of this species is covered with prickles. Coloration of the fish is brown with many irregular pale spots dorsally. Ventrally it is white. It grows to a maximum length of 20 cm.

Harmful organs and harm: It is poisonous to eat as it contains tetrodotoxin like other members of the genus *Takifugu* (https://www.fishbase.in/summary/24268)

Fish samples collected from Jeju and Tongyeong (South Korea) showed toxicity in their internal organs such as skin, liver, ovary, fin, testis and muscles. A toxin level of 79 MU/g was recorded in the muscle; 158-365 MU/g in the fin, intestine, testis, and gallbladder; and 1,147-2,406 MU/gram in the skin, liver, and ovary. Based on these findings, special attention may be required for the users when the fish of this locality is intended for human consumption (Kim et al., 2009).

Takifugu rubripes **(=** *Fugu rubripes* **)**

Image credit: Totti , Wikipedia

Common name: Japanese pufferfish, Tiger puffer

Global distribution: Hokkaido, Japan.

Ecology: It is a demersal and non-migratory species dwelling in freshwater, brackish water and marine habitats

Description/ Biology: Body of this species is covered with prickles. A large round black blotch edged with a white line is present on side just behind pectoral fin. It grows to a maximum length of 80 cm.

Harmful organs and harm: It is poisonous to eat as it contains tetrodotoxin like other members of the genus Takifugu. Its toxin is highly concentrated in liver and ovaries and is slightly present in the intestines and flesh of this species (https://www.fishbase.de/summary/Takifugu-rubripes.html)

Tetractenos glaber

Image credit: Third Silence Nature Photography, Wikipedia

Common name: Smooth Toadfish

Global distribution: Throughout southeastern Australia.

Ecology: It is a demersal species dwelling in freshwater, brackish water and marine habitats

Description/ Biology: It is a small pale yellowish to greenish pufferfish, with dark brown irregular spots and blotches. Belly is white. These spots become fewer on the lower sides. It has a moderately long and robust body which is covered with minute embedded spines. It grows to a maximum length of only 16 cm.

Harmful organs and harm: It is a very poisonous species and should not be eaten (https://fishesofaustralia.net.au/home/species/878). Its toxicity is due to tetrodotoxin, which is concentrated mainly in the liver, ovaries, intestines and skin. According to very earlier reports , this species was responsible for the deaths of the wife and two children of Captain Bell of New Town near Hobart (city of the Australian island state of Tasmania) during 1831. Its poison is a powerful sedative, producing loss of speech, vision and the power of the voluntary muscles, and ultimately complete deprivation of nervous power and death (https://en.wikipedia.org/wiki/Smooth_toadfisha).

Tetraodon lineatus

Image credit: Earedien , Wikipedia

Common name: Globe fish, Nile pufferfish, Fahaka puffer

Global distribution: Africa: Nile, Chad, Senegal, Gambia, Geba, Volta River basins and Lake Turkana in Kenya

Ecology: This demersal species lives in both freshwater and brackish-water; large rivers, open water, and weed beds.

Description/ Biology: In this species, its skin is scaleless, but head and body are covered with small spines except on snout and caudal portion. Teeth form beak-like structure. There are two pairs of non-perforated nasal tentacles. Dorsal and anal fins are short and are placed far back on body. Pectoral fins are well developed and there are no pelvic fins. Caudal fin is rounded. Longitudinal stripes are seen all along sides of body especially in adults. It grows to maximum length of 43 cm and weight of 1.5 kg.

Harmful organs and harm: Like all pufferfishes, this species also carries tetrodotoxin and therefore it is not fit for consumption (https://en.wikipedia.org/wiki/Fahaka_pufferfish)

Tetraodon mbu

Image credit: Tiia Monto , Wikipedia

Common name: Giant puffer, giant mbu puffer

Global distribution: Republic of Congo, Democratic Republic of Congo, Tanzania, Zambia, Burundi, Cameroon.

Ecology:; It inhabits rivers and freshwater lakes; and in brackish-water

Description/ Biology: It is a very large growing puffer and it has a maximum length of 75 cm. It has an elongated body which is mainly yellow/cream in colour with mottled brown markings on the head and top of the body. Belly color varies from cream to clear yellow. Caudal in is long and yellow in color. Eyes are orange.

Harmful organs and harm: This species is known to contain the tetrodotoxin and therefore it is unfit for human consumption (https://www.ch.ic.ac.uk/local/projects/assavathorn/Tetrodotoxin.html)

5.2. Amphibians

The amphibians such as toads, frogs, newts, and salamanders have been reported to produce potent toxins such as digoxin, tryptamines, and tetrodotoxin which may cause several symptoms viz., dizziness, irregular heart rhythm , cardiac arrest, and paralysis. Further, toads and pet frogs are also known to spread salmonellosis, an infection due to the bacteria Salmonella to humans. The symptoms associated with salmonellopsis include include diarrhea, fever, chills, and abdominal pain.

***Bufo* toad secretions**: The most important poisonous components of the *Bufo* toad secretions are cardioactive steroids , catecholamines, tryptamines ((https://www.poison.org/articles/poisoning-from-amphibians-204#:~:text=Some%20of%20the%20potent%20toxins,to%20spread%20Salmonella%20to%20humans). The optimal uses of cardioactive steroids and catecholamines may be of some value in pharmacological applications. For example, the steroid, digoxin is used in microgram amounts to treat heart failure and irregular rates and rhythms of the heart. Similarly, the catecholamines such as epinephrine (adrenaline), norepinephrine, and dopamine have been reported to regulate the human stress or "fight-or-flight" responses. On the other hand, the tryptamines possess adverse effects. The tryptamine component, 5-MEO-DMT or O-methyl-bufotenin which is used in in South America as an entheogen i.e for inducing alterations cognition, behavior mood or consciousness. If and when the eyes or nose of humans are affected by these secretions, they may cause severe irritation, pain, and tissue damage. Unintentional swallowing of these secretions may lead to numbness of the mouth and throat as well as severe effects on the heart which are all due to the digoxin-like compounds and catecholamines mentioned above. Similar affects may also occur after the absorption of secretions through the skin

Fire-bellied toad secretions : Contact of the eyes with the secretions of fire-bellied toad may result in the swelling of the eyelids, visual changes, intense pain and infection. Skin contact may lead to itching and hives.

Secretions of newts and salamanders : The tetrodotoxin (TTX) secreted by the skin, muscles, and internal organs of the newts and salamanders are believed to be the most dangerous toxins known to man. It acts on the nervous system causing muscle paralysis. Further, ingestion of TTX may cause numbness of the tongue, lip, face, arms, and legs. This may be followed by muscle weakness , dizziness and drooling. Death may result if there is no adequate and intime treatment .

Risk: Innocent children and pregnant woman are believed to be at rish when they are exposed to these amphibian secretions. People over 65 years of age, and those who are immunocompromised are advised to be very careful in handling these poisonous amphibians in order to avoid death.

Treatment: Decontamination i.e. removing the toxin by washing, rinsing, flushing, or irrigating with water is the only treatment that can be done at home. It is advisable not to vomit. Though there is an antidote this can only be supportive for the poisoned patients (https://www.poison.org/articles/poisoning-from-amphibians-204#:~:text=Some%20of%20the%20potent%20toxins,to%20spread%20Salmonella%20to%20humans)

5.2.1. Toads

Anaxyrus fowleri (= Bufo fowleri)

Image credit: De Aneta Kaluzna - Trabajo propio, Wikipedia

Class: Amphibia Order: Anura Family Bufonidae

Common name: Fowler's toad

Global distribution: It is native to North America: Eastern United States and Canada

Habitat: It lives near temporary or permanent waters

Description: It is brown, olive green and brownish red in color with dark warty spots. It has a short, rounded snout and a round body. Smaller cranial ridges are also seen. Adult toads are with a pale stripe on their back. Belly is uniformly whitish except for one dark spot. Males have a darker throat than females. These toads have a maximum length of about 10 cm.

Harmful organs and Harm: The skin extracts of this species have been reported to contain biogenic amines such as serotonin, N -m ethyl serotonin, bufotenine, bufotenidine and dehydrobufotenine; and bufadienolides such as arenobufagin, bufalin and gam abufotalin. Among these compounds, the bufadienolides have caused human deaths at higher concentrations (Merovich, 2005).

Atelopus spumarius

Image credit: Mauricio Rivera Correa, Wikipedia

Common name: Pebas stubfoot toad, harlequin toad

Global distribution: Brazil, Colombia, Ecuador, French Guiana, Guyana, Peru, Suriname

Habitat: It is found mainly along the banks of streams

Physical Description: Males 26-29 mm and females 31-39 mm in Length. Body of this species is flattened. Dorsal side has a net-like pattern of green on a brown background. Much of the ventral surface is white. Hands, feet, and posterior surface of the belly are all red. While males grow to a maximum size of about 3 cm, females 4 cm.

Harmful organs and harm: The skin of this species of this genus has been reported to possess the neurotoxin, tetrodotoxin (Kim et al., 1975)

Atelopus zeteki

Image credit: Wikipedia

Common name: Cerro Campana stubfoot toad , Panamanian golden frog

Global distribution: It is endemic to Panama

Ecology: It lives in fast-moving streams

Description/ Biology: Skin colour of this species varies from light yellow-green to bright gold. However, some individuals may exhibit black spots on their back and legs. Females are generally larger than males. While females grow to a maximum length of 6 cm with a weight of 15 g, males are with 5 cm and 12 g respectively

Harmful organs and harm: It is one of the most toxic species of *Atelopus* and its skin contains a toxin, Zetekitoxin which is a hypotensive substance, with the ability to cause cardiac arrhythmias and ventricular failure in dogs and cats (Toledo and Jared,1995). Further, the skin of a single frog contains enough toxins to kill 1,200/20g sized mice (https://en.wikipedia.org/wiki/Atelopus) . It is also reported that this frog has a variety of toxins, including steroidal bufadienolides and guanidinium alkaloids of the class, tetrodotoxin . Its alkaloids viz. zetekitoxin AB, have been found to be a blocker of voltage-dependent sodium channels and these toxins may affect the nerve cells of anyone who comes in contact with it. High doses can be fatal in 20 or 30 min. Death is invariably preceded by clonic convulsions until the functions of the circulatory and respiratory systems stop (https://en.wikipedia.org/wiki/Atelopus)

Bufo bufo (= *Bufo vulgaris*)

Image credit: Karamell, Wikipedia

Common name: Common toad, European toad

Global distribution: It is found throughout Europe (excepting Ireland, Iceland, and some Mediterranean islands), , western part of North Asia, and Northwest Africa.

Habitat: Breeding of this species takes place in ponds

Physical Description: It has more warty skin, and very striking coppery eyes.

Harmful organs and harm: The parotid glands of *Bufo* spp. have been reported to release toxic substances when these toads are threatened. These toxic substances such as dopamine, norepinephrine, epinephrine, serotonin, bufotenine, bufogenin, bufotoxins, and indolealkylamines have shown cytotoxic, cardiotoxic and hemolytic activity.(Wikipedia) . Apart from these substances, the parotoid extract of this species had harmful peptides such as serine proteases, muscle creatine kinase, phospholipid hydroperoxide glutathione peroxidase, cytotoxic T-lymphocyte protein, etc. These peptides have been reported to show cytotoxic, cardioinhibitory and the paralytic (myo- and neurotropic) properties (Kowalski et al., 2018)

Bufo gargarizans

Image credit: Visviva , Wikipedia

Common name: Asian toad; European toad

Global distribution: East Asia, China, Russia and Korean Peninsula

Ecology: Freshwater habitats

Description/ Biology: The average body size of females of this species was significantly larger than that of males (females: 10.2 cm; males: 9.3 cm)

Harmful organs and harm: The toxic substances of this species are bufadienolide compounds which are of two types viz. bufogenins and bufotoxins. Among them, bufagenins are more toxic than bufotoxins. More than 50 different bufagenin compounds have been identified in this Asian toad. Among them, gamabufotalin, telocinobufagin, bufotalin , cinobufotalin , bufalin , cinobufagin and resibufogenin are the dominant components, representing more than 80% of the total amount of bufagenins (Cao et al., 2019). The oral mucosa of this species caused lethality of other animals on contact. Further, cats and dogs which bite these toads salivated profusely with vomit reflexes from a period which varies from 8 to 12 hr. In cats, the poisoning associated with the swallowing these toads produced acute abdominal pain, hyperthermia and wobbly gait (Toledo and Jared,1995; https://www.britannica.com/science/steroid/Toad-poisons#ref100185)

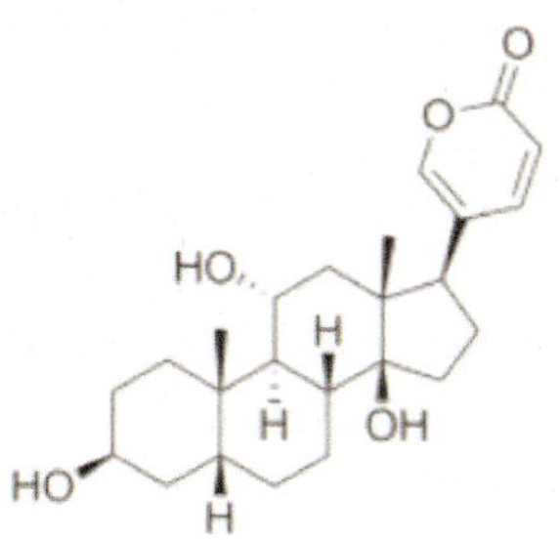
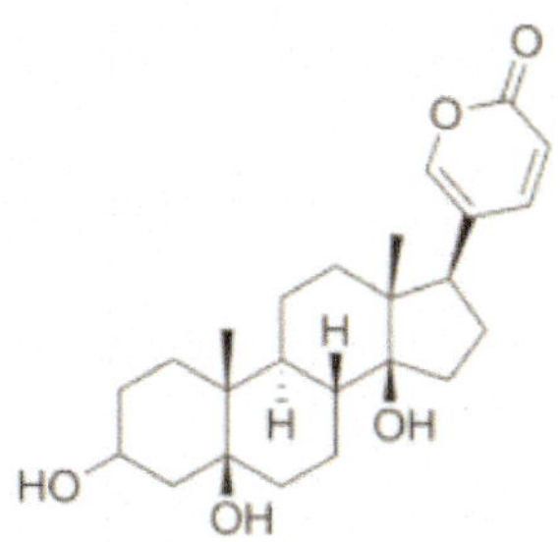
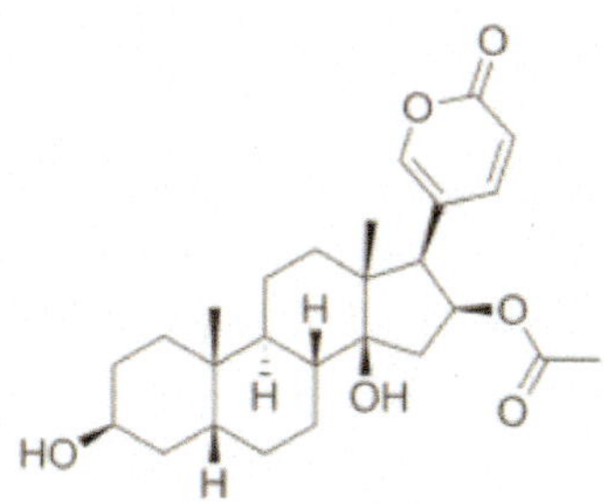

Gamabufotalin Telocinobufagin Bufotalin

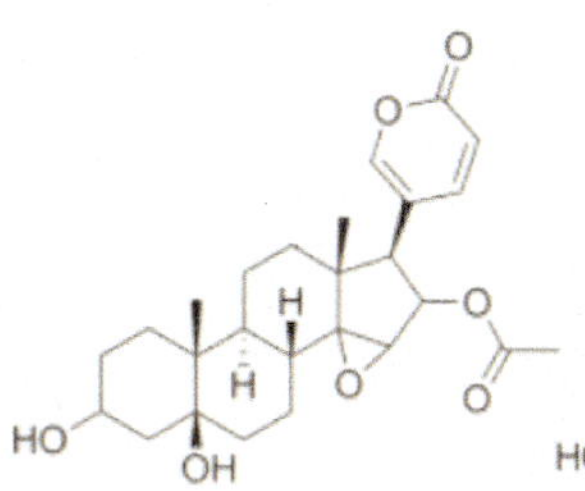
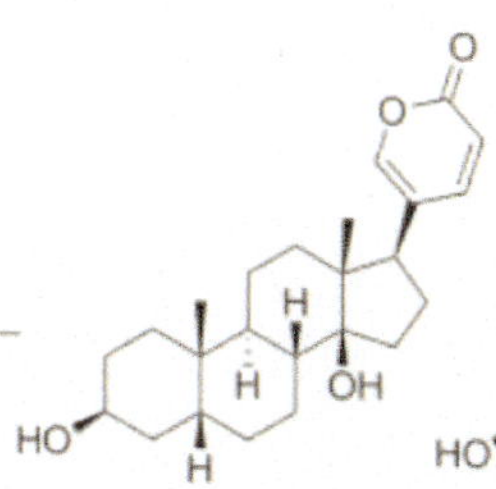
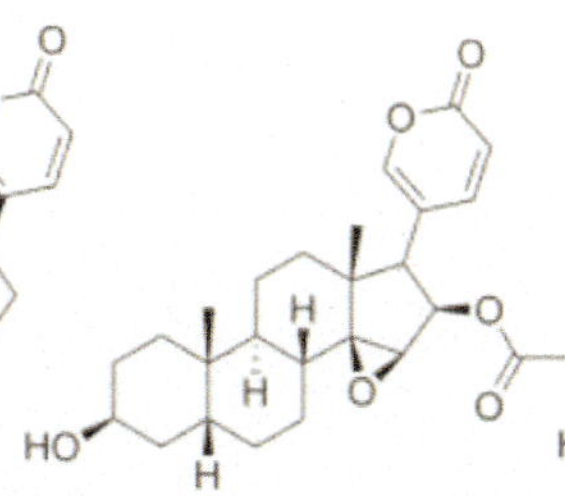
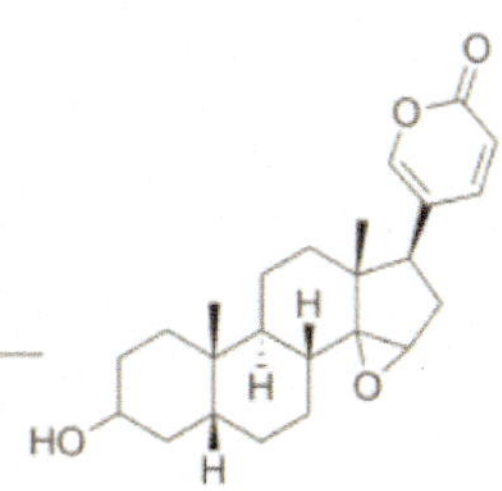

Cinobufotalin Bufalin Cinobufagin Bufogenins of Resibufogenin
Bufo gargarizans

Duttaphrynus melanostictus (= *Bufo melanostictus*)

Image credit: Wikipedia

Common name: Asian common toad, Asian black-spined toad, Asian toad, black-spectacled toad, common Sunda toad, and Javanese toad

Global distribution: South and Southeast Asia

Habitat: It breeds in still and slow-flowing rivers, ponds and pools

Physical Description: In this species, several bony ridges are seen on the top of head, , along the edge of the snout, in front of the eye, above the eye, and behind the eye. Snout is short and blunt. Ear drum is very distinct. Its first finger is often longer than the second finger and the toes are half webbed. A warty tubercle is present just before the junction of the thigh and shank and two moderate ones are on the shank. Dorsal side is covered with spiny warts and the parotoids are prominent and elongated. While the dorsum is yellowish or brownish, spines and ridges are black. It is spotted ventrally. Males possess a subgular vocal sac and black pads on the inner fingers. It grows to a maximum length of 20 cm .

Harmful organs and harm: The skin extract of this species has hypotensive, cardiotoxic and neurotoxic properties and produced death in experimental animals (Das et al., 2001).

Incilius alvarius (= *Bufo alvarius*)

Image credit: Wikipedia

Family: Bufonidae

Common name: Colorado river toad, Sonoran toad

Global distribution: Southern Arizona, southern California and southern New Mexico

Habitat: It is found in streams, canals and drainage ditches,

Physical Description: This species has a smooth, leathery skin and its body is olive green or mottled brown in color. A bulging kidney-shaped parotoid gland is present Just behind the large golden eye . A white wart is seen at the corner of the mouth and there are white glands on the legs. It grows to a maximum length of 19 cm and is said to be the largest native toad in the United States.

Harmful organs and harm: Toxins secreted by species has been reported to cause severe cardiac problems, including abnormal heart rhythms and neurological and gastrointestinal problems. Exposure to the toxin slowed down the heart rate and caused the heart to beat erratically in cats and dogs leading to death (https://www.dixieanimalhospital.com/blog/13014-beware-toads-that-can-poison-cats-and-dogs).

Toledo and Jared (1995) reported that the skin and parotoid glands of this species produce a powerful hallucinogen called 0-methylbufotenine

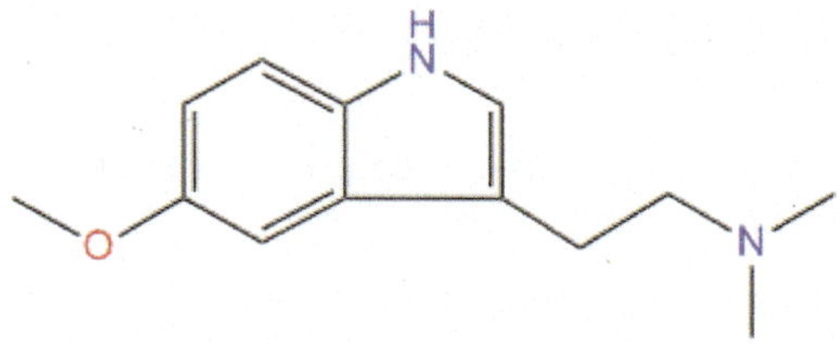

0-methylbufotenine

This species is known to secrete large amounts of hallucinogen, 5-methoxy-N,N-dimethyltryptamine (5-MeO-DMT). The venom of this species hs been reported to be toxic when consumed orally (Davis and Weil, 1992)

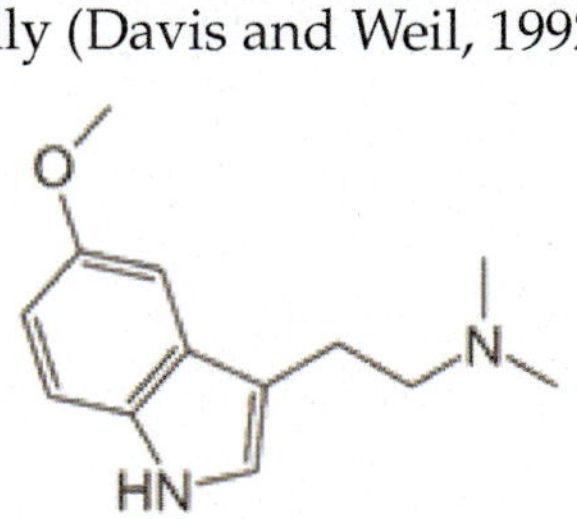

5-MeO-DMT

Phrynoidis aspera (= *Bufo asper*)

Image credit: Wikimedia Commons

Family: Bufonidae

Common name: Asian giant toad

Global distribution: Indonesia, Malaysia, Thailand, Myanmar , Vietnam and Borneo

Habitat: It lives all along the banks of small to medium-sized streams and large rivers

Physical Description: It has a large and stout body. Skin is covered with warts or tubercles. Head is broad and blunt, without bony crests. It has an ovoid parotoid gland connected to the supraorbital ridge. Tympanum is visible. Hands and feet are spinous. Its fourth toe is the longest, and all the toes except the fourth are completely webbed. Males possess nuptial pads at the base of the first finger. Body color is dark brown, gray or black with black spots ventrally. Throats of males are blackish.

Harmful organs and harm: The smoke due to the burning of this frog is believed to induce drowsiness and it is also toxic to the respiratory system of mice and rats (Toledo and Jared,1995).

Skin toxins of this species have been reported to induce prostration, and convulsions in mice at a dose of 100 mg/mouse. The major toxic components of its skin extracts are bufotalin (a bufadienolide), resibufogenin and other bufadienolides and bufotoxins. However, this species is consumed for food in Sabah and peninsular Malaysia (https://amphibiaweb.org/species/110)

Rhinella diptycha (= *Bufo paracnemis*)

Image courtesy: Transpantaneira, Poconé, Mato Grosso, Wikipedia

Order: Anura Family: Bufonidae

Common name: Cope's toad, Schneider's toad, cururu toad, or rococo toad

Global distribution: Argentina, Paraguay, Uruguay, Bolivia, and Brazil

Habitat: Its common habitats are freshwater marshes. It breeds in permanent and temporary ponds

Physical Description: It is a large toad measuring a length of 25 cm for females and 18 cm for males.

Harmful organs and harm: The powdered form of the cutaneous secretion of this species showed toxic characteristics. The symptoms of this poisoning include inflammation of mouth and throat, hiccups, vomiting, dysentery, fainting and convulsions (Toledo and Jared,1995).

Rhinella jimi

Image credit: Mike Peel , Wikipedia

Family: Bufonidae

Common name: Not reported

Global distribution: Entire northeastern region of Brazil

Habitat: Its breeding takes place in permanent and temporary ponds.

Description: It is a stout and large toad measuring an average snout–vent length of 17.1 cm in males and 13.4 cm in females. Tympanum is distinct. Head has many cranial crests. Males possess moderately strong arms, while they are slim in the case of females. Fingers have no webbing. Legs are short and robust. Toes are fringed and are with basal webbing. Its dorsal skin is covered by several tubercles of different sizes. Males are provided with keratinized spines on their dorsum, flanks, and upper surface of limbs. Belly is lighter beige than its dorsal side. Head is black.

Harmful organs and Harm: It is a venomous frog. Kidneys of the venom- treated chicken showed hemorrhage in the interstitium of the renal cortex and vacuolar degeneration in the renal tubular epithelium. Similar observations were also made in poisoned humans (Gadelha et al., 2014).

Rhinella marina **(=** ***Bufo marinus*****)**

Image credit: Benjamint444 , Wikipedia

Family: Bufonidae

Common name: Cane toad or giant toad

Global distribution: This species is native to extreme southern Texas through Central South America.

Ecology: These toads are generally found in terrestrial habitats in areas with permanent nearby water bodies including man-made temporary pools, ponds, and canals for the purpose of breeding

Description/ Biology: Adult toads of this species may reach a size of more than 20 cm. Based on the texture of the back skin and coloration ,males and females are distinguished. While females have smooth, mottled, brown and white backs, males are more yellow with rough backs.

Harmful organs and harm: All stages of the cane toad's life cycle viz. eggs, tadpoles, toadlets and adult toads are poisonous. Adults of this species are most toxic and it is largely due to the amount of poison contained in their enlarged parotoid glands present on their shoulders (https://edis.ifas.ufl.edu/pdffiles/UW/UW43200.pdf). The toxins secreted by this species can cause severe cardiac issues including abnormal heart rhythms as well as neurological and gastrointestinal problems in mammals. Exposure to the toxin slows down the heart rate and causes the heart to beat erratically. Death may result in pet animals (https://www.dixieanimalhospital.com/blog/13014-beware-toads-that-can-poison-cats-and-dogs). Toledo and Jared (1995) stated that small doses of the toxin of this species may cause cardiac arrhythmia, pulmonary edema, hypertension, convulsions and death in dogs.

Gowda et al. (2003) reported that this toad's venom a constituent of an aphrodisiac, was found to be responsible for the development of several clinical manifestations and death in any patient affected by this venom. Digoxin-specific antibody (Fab) fragments have been reported to be beneficial in the treatment of toad venom poisoning.

Colostethus inguinalis (= *Colostethus panamansis*)

Image courtesy: Wikipedia

Order: Anura Family Dendrobatidae

Common name: Common rocket toad

Global distribution: Endemic to northwestern and northcentral Colombia

Habitat: It is living along streams

Physical Description: It is a moderate-sized dendrobatid toad with a maximum length of 3.0 cm in females and 2.7 cm in males. In adult males, finger III is swollen in adult males. Throat and) chest and anterior belly are black colored in adult males and are white or faintly pigmented gray or brown in adult females. A pale oblique lateral stripe which is extending from groin midway to eye is present. A pale ventrolateral stripe is also present.

Harmful organs and harm: This toad is toxic due to its tetrodotoxin content of its skin. Small amounts of anhydrotetrodotoxin and 4-epietrodotoxin were also present in this species (Daly et al., 1994)

5.2.2. Frogs

Dendrobates auratus

Image credit: Adrian Pingstone, Wikipedia

Order: Anura Family: Dendrobatidae

Common name: Green and black dart-poison frog, green-and-black poison arrow frog and green poison frog

Global distribution: Central and South America: from Nicaragua and Costa Rica to southeastern Brazil and Bolivia

Habitat: It prefers locations near small streams or pools.

Physical Description: This species has many color variants. Most of such variants are black and some may be green or light blue, with the black in bands or spots. Adults are about 4 cm long and they have a fused head and trunk with no tail. Poison glands are located throughout the surface of their body.

Harmful organs and harm: The skin of this frog is highly toxic and contact of this animal can be dangerous. It is reported that the very small amount of the poison of this frog is enough to make a human heart stop beating (https://animaldiversity.org/accounts/Dendrobates_auratus/)

The venom of *Dendrobates* spp. has bbeen reported to contains various types of piperidine alkaloids such as histrionicotoxins, pumiliotoxin C and the gephyrotoxins all of which act on ion channel associated with the acetylcholine receptor of the neuromuscular junction. Further, they interact with the acetylcholine receptor, blocking the passage of sodium and potassium ions. This blockage is known to prevent the acetylcholine from triggering muscle contraction. Furthermore , the pumiliotoxins, when injected subcutaneously in mice, causde problems with locomotion, clonic convulsions and the death of the animals (Toledo and Jared,1995).

Dendrobates leucomelas

Image credit: Arpingstone , Wikipedia

Common name: Yellow headed poison frog

Global distribution: Northern South America: Venezuela, Brazil, Guyana and Colombia

Habitat: It is associated with freshwater habitats.

Physical Description: This species is identified by its yellow and black bands across the body. During growth, these black bands break off into spots. Females of this species are generally larger and more robust than males. It grows to a maximum length of 5 cm and it has an average weight of only 3 g.

Harmful organs and harm: The skin toxins of this species have been reported to be very intense, and are capable of killing or injuring humans if these frogs are not carefully handled (https://animaldiversity.org/accounts/Dendrobates_leucomelas/).

Dendrobates tinctorius

Image credit: Wikipedia

Common name: Dyeing dart frog or dyeing poison dart frog

Global distribution: Throughout the eastern portion of the Guiana Shield and Venezuela; Guyana, Suriname, Brazil, and French Guiana

Habitat: It lives near streams

Description: The length of this species may exceed 7 cm. Body of this species is black, with an irregular pattern of yellow or white stripes which run all along the back, flanks, chest, head, and belly. In some morphs of this species body color may be blue, yellow or white. While legs are typically with small black dots, their color largely varies in different morphs from pale blue, to navy blue or royal purple. Males are generally smaller and more slender than females, and they have larger toe discs. Toe discs of female frogs are circular while those of males are heart-shaped. Further, females have arched backs against curved ones in males.

Harmful organs and Harm: This species has been reported to be highly toxic due to its pumiliotoxin and allopumiliotoxin. By ingestion, these toxins cause pain, cramping, and stiffness and other serious effects on humans including death (https://www.treehugger.com/beautiful-deadly-frogs-4859270)

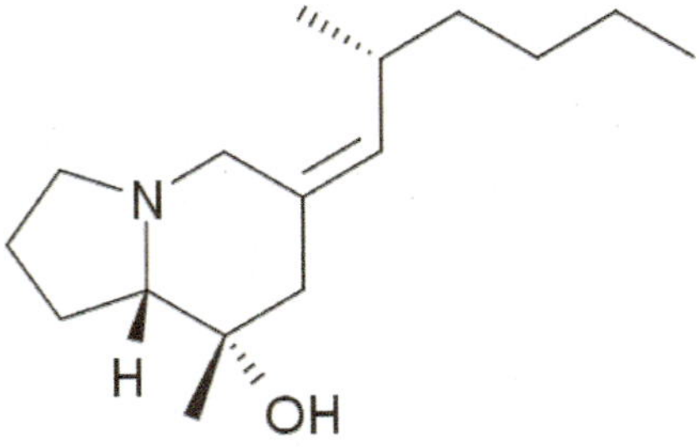

Pumiliotoxin

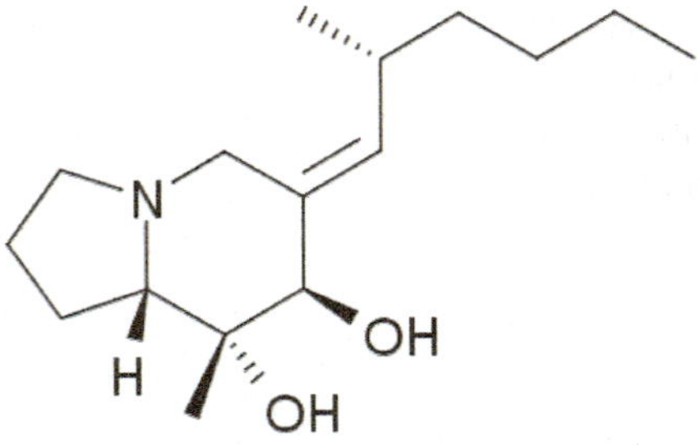

Allopumiliotoxin

Epipedobates tricolor

Image credit: Holger Krisp , Wikipedia

Common name: Phantasmal poison frog or phantasmal poison-arrow frog

Global distribution: It is endemic to Ecuador

Habitat: It lives near streams, and in wetlands

Description: This tiny, beautiful species is in endangered category. It has radiant colors, a wide head, truncated snout and smooth skin. First finger is longer than the second, and all the digits are partially webbed. Dorsal surface is green or yellow with several longitudinal stripes. The phantasmal poison frog is not only beautiful, it is also exceptionally tiny. It grows to only 4 cm in length.

Harmful organs and Harm: This frog has been reported to carry enough poison to kill an adult human. It has an alkaloid toxin called epibatidine which is believed to be a non-addictive painkiller more powerful than morphine. While this toxin has promising, bioactivity, scientists have determined that this compound may also be too toxic to humans (https://www.treehugger.com/beautiful-deadly-frogs-4859270).

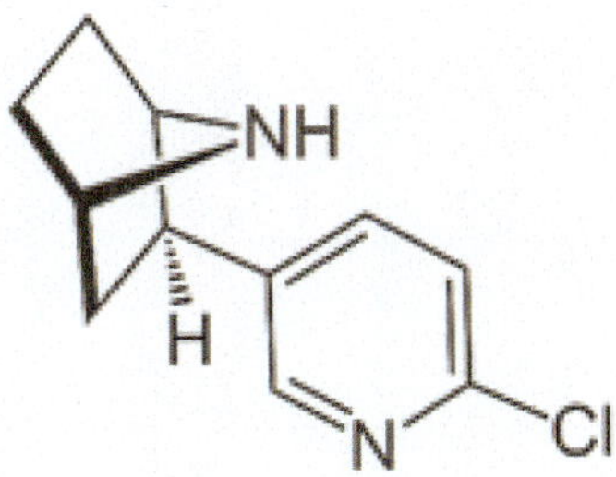

Epibatidine

Oophaga grnulifera (= *Dendrobates granuliferus*)

Image credit: Wikipedia

Family: Dendrobatida

Common name: Granular poison frog

Global distribution: It is found in Costa Rica and Panama

Habitat: It dwells near streams for breeding purpose

Description: It is a small frog with slender limbs. Its skin is finely granulated and it has bright orange head, body and upper arms. Belly, legs and lower arms are bluish-green . Its other morph bears olive green or bright red body . Calling males may have a black vocal pouch. It grows to a maximum length of only 2 cm.

Harmful organs and Harm: The toxic skin secretion of this species makes it unsuitable for ingestion (https://amphibiaweb.org/cgi/amphib_query?where-genus=Oophaga&where-species=granulifera&account=amphibiaweb)

Oophaga histrionica

Image credit: Esteban Alzate, Wikipedia

Common name: Harlequin poison frog or harlequin poison-dart frog

Global distribution: It is endemic to the Chocó region of western Colombia.

Habitat: It lives near freshwater bodies to facilitate breeding

Description: This species has a variety of color morphs and its base color is bright orange, with a webbing of black on the entire body. This web pattern may vary from small lines to big lines, speckled, or completely black with a few spots. It has smooth skin, and no tarsal tubercle. Males are generally of the same size or larger than the females. It reaches a maximum size of 4 cm.

Harmful organs and Harm: This species has been reported to produce a cardiotoxin known as histrionicotoxin which is a potent non-competitive antagonist of nicotinic acetylcholine receptor. This toxin blocks action potentials and slows neural function. (http://www.t3db.ca/toxins/T3D2528)

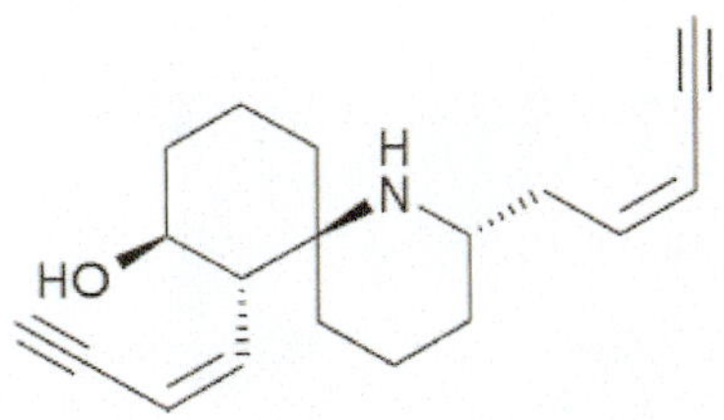

Histrionicotoxin

Oophaga pumillo (= *Dendrobates pumillo*)

Image credit: Wikipedia

Common name: Strawberry poison frog or strawberry poison-dart frog

Global distribution: Central America; Eastern central Nicaragua and Northwestern Panama

Habitat: It needs moist, terrestrial habitats with abundant water-filled plants for reproduction

Description: This species is known for its widespread variation in coloration and about 30 color morphs have so far been recorded. Color of these morphs varies from red to blue or green with black spots. It is a small, slender frog with bilateral symmetry. Other characteristic features of this species include : presence of four, un-webbed digits on each hand and foot; quite compact body and large and dark eyes; and very moist skin which gives this frog a somewhat glossy appearance in bright light. This small frog reaches a maximum length of only 2.4 cm.

Harmful organs and Harm: It is the most toxic member of its genus (Wikipedia). Like other dart frogs, this species derives its toxicity from its diet of ants and termites. In captivity, these frogs have been reported to lose its toxin (https://www.treehugger.com/beautiful-deadly-frogs-4859270).

Oophaga speciosa

Image credit: Wikimedia Commons

Family: Dendrobatidae

Common name: Splendid poison frog, splendid poison-arrow frog, or Panamanian red poison dart frog

Global distribution: Panama

Habitat: It lives near freshwater bodies for breeding

Description: It is a small poison frog growing to a maximum length of only 3 cm. Its dorsum which has a smooth skin is deep red, with faint black, irregular venations. Its inner tarsal fold reduced to a tubercle. Tympanum is round and its diameter is slightly greater than one half the diameter of its eye.

Harmful organs and Harm: This species has been reported to produce a cardiotoxin known as histrionicotoxin which is a potent non-competitive antagonist of nicotinic acetylcholine receptor. This toxin blocks action potentials and slows neural function. (http://www.t3db.ca/toxins/T3D2528).

Phyllobates aurotaenia

Image credit: TomR (Thomas Ruedas) , Wikipedia

Common name: Kokoe poison frog

Global distribution: Central and South America

Habitat: In this species, male frogs takes the tadpoles to slow-flowing water after the eggs hatch.

Description: The frogs of this species are characterized by: black dorsums which are sometimes covered by orange coloration with green, yellow or brownish gold dorsolateral stripes. Belly is black with blue or green dots. Two forms of this species viz. a smaller, large-striped form and a larger, small-striped form have been reported. Skin is slightly granular dorsally and belly and limbs are smooth. First finger is longer than the second and all finger discs are either narrow or moderately expanded. Toes are without web. Females are slightly larger than males and they reach a maximum length of 3.5 cm and 3.2 cm respectively.

Harmful organs and Harm: The cutaneous granular glands which are scattered throughout these frogs' bodies have been reported to secrete the steroidal alkaloids viz. batrachotoxin, homobatrachotoxin, and batrachotoxinin A, which cause depolarization of nerves and muscles, cardiac arrhythmias, and cardiac failure if ingested and skin numbness when these toxins are encountered externally. It has also been reported that its batrachotoxin which is derived through various diets of this species such as melyrid beetles, millipedes, and flies may lead to paralysis and often death even at a small dose of 100 µg (https://amphibiaweb.org/species/1704).

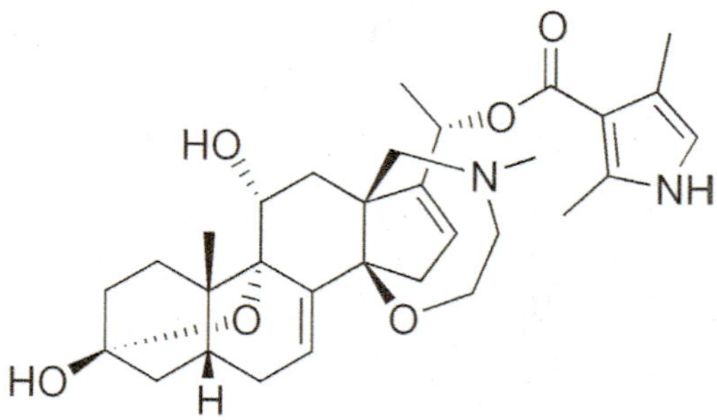

Batrachotoxin

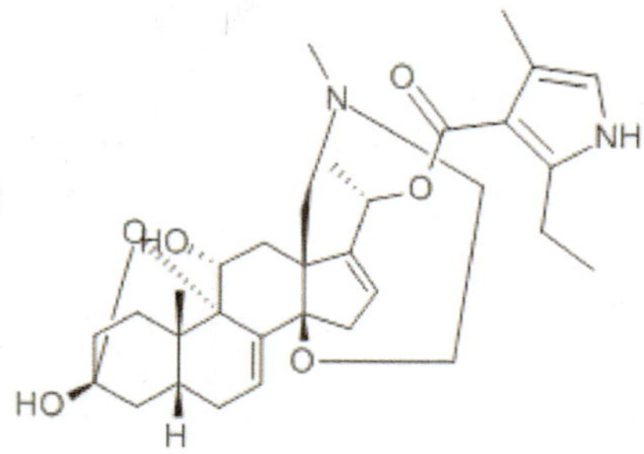

Homobatrachotoxin

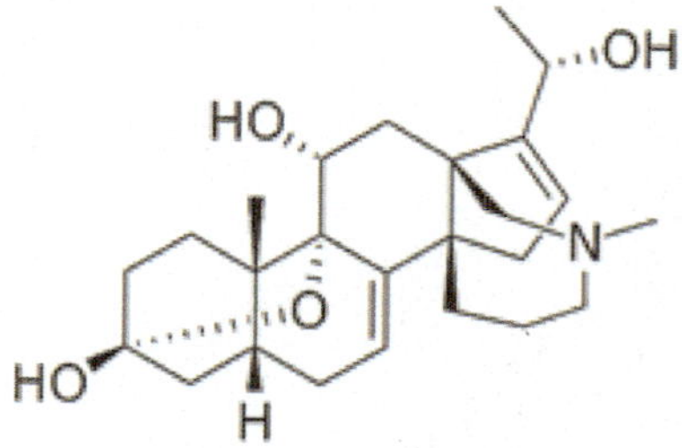

Batrachotoxinin A

Phyllobates bicolor

Image credit: Luis Miguel Bugallo Sánchez, Wikipedia

Common name: Black-legged poison frog, bicolored dart frog

Global distribution: Colombia

Habitat: It lives in freshwater bodies like drainages and headwaters

Description: It is a small frog, with males reaching a maximum length of about 4 cm and female, 4.3 cm. Skin is smooth. Its color patterning is very striking. While its dorsum and sides are a uniform golden yellow or orange, ventrum is completely black, or washed with light orange or bluish green. A black patch may be present on the throat in some specimens. Tympanum is yellow-orange and black. Tip of the snout is black in some individuals. Irises are black or reddish-brown. Both premaxillary and maxillary teeth are present. First finger is longer than the second, and digits show expanded discs. Webbing is absent in toes.

Harmful organs and Harm: It is a highly toxic animal due to its batrachotoxin, the amount of which has been reported to be 17-56 µg per frog. It is one of the few frogs confirmed for its human fatalities. And 150 µg of its poison would be enough to kill an adult human (https://amphibiaweb.org/cgi/amphib_query?where-genus=Phyllobates&where-species=bicolor)

Phyllobates lugubris

Image credit: Wikipedia

Common name: Lovely poison frog or lovely poison-arrow frog

Global distribution: Central America; Costa Rica, Panama

Habitat: Adult frogs are often found near rocky regions of forest streams. It also prefers to live near slow-moving waters.

Description: It is a small frog reaching a maximum length of 2.4 cm in females and 2.1 cm in males. Skin is slightly granular on the dorsum, dorsal surface of the limbs, belly, and the ventral surface of the thighs. However, skin is smooth on the ventrum and ventral surface of the limbs. Head is longer with a rounded and truncate snout. Both maxillary and premaxillary teeth are present. First finger is longer than the second and toes are without web. Males possess a thick, dark nuptial pad in each thumb, and thicker forearms. Maxillary teeth are present. Eyes are large with dark brown irises. Dorsum is black with a pair of pale yellow or orange dorsolateral stripes which run along the sides of the dorsum from the hindlimbs to the snout. Limbs are mottled with yellow–greenish yellow. Ventral surfaces are mottled with black pigments giving a solid black appearance.

Harmful organs and Harm: It is a toxic frog due to its batrachotoxins of skin secretions. These toxins are by this animal acquired from its dietary sources such as ants, mites or beetles (https://amphibiaweb.org/species/1706)

Phyllobates terribilis

Image credit: Wilfried Berns , Wikipedia

Family: Dendrobatidae

Common name: Golden poison frog, golden frog, golden poison arrow frog or golden dart frog

Global distribution: It is endemic to Pacific coast of Colombia

Ecology: It lives near smaller streams and its breeding is in permanent pools

Description/ Biology: It is a small frog with adult females having a maximum length of 4.7 cm, and adult males reaching a size of 4.5 cm. Snout is rounded in lateral profile, and is bluntly rounded to truncate dorsally. Hands and feet lack webbing. Skin is smooth to fairly rugose or finely granular, Teeth are present on the maxillary arch. Males possess a shallow subgular vocal sac at the base of the throat, and well-developed paired vocal slits on the floor of the mouth. Dorsal coloration is bright golden yellow or pale metallic green. Some individuals may have a deep orange or pale greenish yellow coloration. Ventrally the color is the same except for the underside of the hands and feet, which are black, and the undersides of the thighs, which are with a black seat patch. Eyes and nares are black.

Harmful organs and harm: It is a lethally toxic frog due to its batrachotoxin, a steroidal alkaloid produced by this frog at 700-1900 µg per frog. The toxin from a single individual has been reported to kill 20,000 mice or 10 humans (Wikipedia). Batrachotoxin is cardio-, myo-, and neurotoxic and it disturbs coordination and equilibrium. It is also said to cause clonic convulsions and cardiac arrhythmia (Toledo and Jared,1995)

Phyllobates vittatus

Image credit: Wikipedia

Common name: Golfodulcean poison frog or Golfodulcean poison-arrow frog

Global distribution: It is endemic to Costa Rica

Habitat: It is a diurnal and terrestrial species associated with streams. It also prefers forested valleys with streams

Description: It is a small frog, with males reaching a size of 2.6 cm and females measuring up to 3.1 cm. Skin of the dorsal surface is slightly granular with tiny bumps. Skin of the ventral surfaces of the thighs and belly are also slightly granular , but it is smooth on the rest of the ventral surface. Both maxillary and premaxillary teeth are seen. Finger I is longer than finger II and toes lack webbing. Coloration of back and head are solid black, albeit in some individuals there may be an interrupted yellow median dorsal stripe. A wide red-orange or orange dorsolateral stripe is seen running from each side of the snout to the base of the thigh. Dorsal surfaces of the limbs display thick blue-green speckling on a black background, while the venter and ventral surfaces of the limbs are marbled with white coloration. Each flank has a pale blue-green stripe which is running ventro-laterally .

Harmful organs and Harm: Like all members of the genus *Phyllobates*, it also has a highly potent neurotoxic alkaloid poisons in its skin. Its poison has been reported to cause severe pain, and even paralysis if a larger dose of the toxin is administered. It acquires its toxin from a species of insect on which it feeds (Wikipedia). According to some researchers , the toxin of this genus viz. batrachotoxins persists in wild-caught frogs of this species even when they are maintained in captivity (https://amphibiaweb.org/species/1708).

Ranitomeya variabilis

Image credit: Tony Hisgett from Birmingham, Wikipedia

Order: Anura Family: Dendrobatidae

Common name: Zimmermann's poison frog or variable poison frog

Global distribution: Colombia, Ecuador, Peru.

Habitat: Its breeding is in pools.

Description: It has an average snout-vent length of 1.7 cm in adult males and 1.8 cm in females. Body coloration of this frog is from yellow-green to a blue-green from anterior to posterior. It has a single nose spot with large dorsal spotting. Limbs are slightly black with small, even spots.

Harmful organs and Harm: It is a toxic frog due to its alkaloid poisons (https://en.wikipedia.org/wiki/Zimmerman%27s_poison_frog).

Ranitomeya reticulata

Image credit: Wikispecies

Common name: Red-backed poison frog or reticulated poison frog

Global distribution: It is native to the Amazon rainforest in Peru and Ecuador

Habitat: Its breeding is in freshwater bodies

Description: It is one of the smaller species of poison dart frogs as males reach a maximum length of 1.2 cm and females have a length of 2 cm. It is also beautifully colored and patterned. It has black legs with a sky-blue mesh pattern, a black abdomen , and a back, the color of which varies from fiery orange to scarlet. As in all arboreal frogs, it possess sucker- like disks on its toes.

Harmful organs and Harm: It has been reported to be a moderately toxic species and its poison is capable of causing serious injuries to humans. It acquires its toxins from its food sources such as ants, mites, and beetles. (https://en.wikipedia.org/wiki/Red-backed_poison_frog)

Dryophytes cinereus (= *Hyla cinerea*)

Image courtesy: Wikimeida Commons

Order: Anura Family: Hylidae

Common name: American green tree frog

Global distribution: Central and south-eastern United States

Habitat: It is commonly found in small ponds, large lakes, marshes, and streams with plentiful floating vegetation, grasses, and cattails

Physical Description: This frog is green, medium-sized, and is growing up to a length 6 cm. Body is green in shades ranging from yellowish-olive to lime green. Small patches of gold or white may also occur on the skin. White, pale yellow, or cream-colored lines may also run from its jaws to groins. It has smooth skin and large toe pads. Abdomen is pale yellow to white. Males possess wrinkled throats and are slightly smaller than females in size .

Harmful organs and harm: Biogenic amines such as histamine and serotonin have been reported from the skin of this species. Histamine has hypotensive properties besides serving as a stimulant of the gastric secretions. It is also involved with hypersensitivity reactions and sensorial cutaneous impulses which are said to cause itching and pain (Toledo and Jared,1995)

Mantella aurantiaca

Image credit: Wikipedia

Order: Anura Family: Mantellidae

Common name: Golden mantilla frog

Global distribution: It lives only in the east-central Madagascar

Habitat: It prefers swampy areas.

Physical Description: It is an attractive little frog and color of dorsum is ranging from bright yellow - orange to red- orange. Bright red flash marks are present on the body. Iris is uniformly black. Belly is somewhat lighter, except red flash mark . It grows to a maximum length of 3.1 cm. especially females. This species does not show any sexual dimorphism, but males are known by its small size (compared to the female) and the presence of femoral glands on the undersides of their thighs. It is a "Critically Endangered" species due to its limited distribution.

Harmful organs and harm: This species has been reported to be toxic and it retains its toxins from the insects it eats . A variety of toxins viz. pumiliotoxin, allopumiliotoxin and homopumiliotoxin alkaloids, as well as pyrrolizidines, indolizidines and quinolizidines has been reported form its skin (https:// amphibiaweb.org/species/4559)

Mantella madagascariensis

Image credit: Wikipedia

Order: Anura Family Mantellidae.

Common name: Malagasy painted mantella, Madagascan mantella, Madagascar golden frog, Malagasy mantella and painted mantella

Global distribution: It is endemic to East-central Madagascar: rom Niagarakely to Ranomafana,

Habitat: It prefers moist montane forest and rivers; and is also found near streams.

Physical Description: It is a tiny but stout frog with variable color and markings. Body is irregularly streaked in black and yellow. Usually it has a blackish body, but its dorsum is commonly black. A green or yellowish rostral stripe runs along the head. Femur and humerus are yellow to green, while tibia, tarsus and foot are reddish-orange. Blotches of yellow-green color extends to flanks. While female have a maximum length of 2.5 cm, males measure 2.2 cm.

Harmful organs and harm: This species has been reported to produce 16 toxic alkaloids in its skin. These alkaloids are derived from its ant diet or through the plant-ant -frog toxin food chain (https://news.mongabay.com/2005/08/study-discovers-why-poison-dart-frogs-are-toxic/).

Pseudophryne corroboree

Image credit: Andrew , Wikipedia

Family: Myobatrachidae

Common name: Southern Corroboree Frog

Global distribution: Its distribution is restricted to the Snowy Mountains Regions of Kosciuszko National Park in New South Wales, Australia.

Habitat: Breeding occurs near shallow pools and seepages

Physical Description: This species is slightly smaller in size and it has its characteristic bright yellow stripes. Ventral surface is boldly marked with black, yellow and white blotches. Adult frogs reach a maximum length of 3.0 cm.

Harmful organs and harm: The poisonous alkaloid viz. pseudo-phrynamine secreted from the skin of this species has been reported to be potentially lethal to mammals if ingested (https://en.wikipedia.org/wiki/Corroboree_frog). This species is unique among its related species as it produces its own poison rather than from its food source.

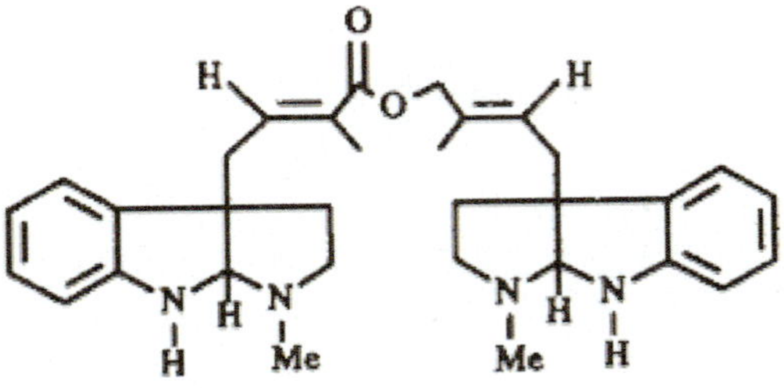

Pseudo-phrynamine

Pseudophryne pengilleyi

Image credit: Wikipedia

Common name: Northern corroboree frog

Global distribution: Sub-alpine regions of the Australian Capital Territory and southern New South Wales

Habitat: This species prefers to breed in bog pools at high altitudes and in shallow seepage pools in gullies

Description: It is a small species of frog reaching up to 3 cm in length. It has a distinctive dorsal greenish-yellow coloration with shiny black stripes which are broken and interconnected. This pattern extends also over the limbs and flanks. Belly is marbled with either black and white or black and yellow. A large flat femoral gland is seen on each hind limb. Inner toe has a single phalanx.

Harmful organs and Harm: It has been reported to produce its own poisonous alkaloid from the skin. This toxin named pseudophrynamine has been described as potentially lethal to mammals if ingested ((https://en.wikipedia.org/wiki/Pseudophryne)

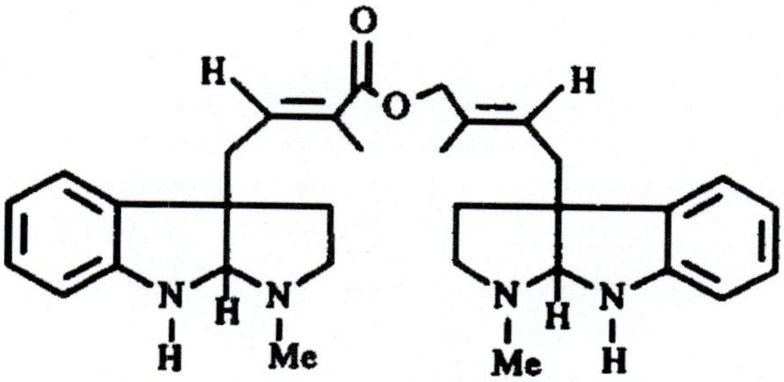

Pseudophrynamine

Phyllomedusa bicolor

Image credit: Jean-Marc Hero , Wikimedia

Family: Phyllomedusidae

Common name: Giant monkey frog , giant leaf frog, and waxy-monkey treefr

Global distribution: Amazon basin , Brazil; Bolivia, Colombia and Peru

Habitat:

Physical Description: Dorsum of this species is green and belly is cream , yellow or white. Its chest, lower and front legs have white spots which are with dark frames. These spots are more on its hind legs and flanks, Fingers which are brown possess large, green discs. A prominent gland is seen extending from its eye to the tympanum. While males attain a maximum length of about 100 mm, females 120 mm.

Harmful organs and harm: The skin secretion (kambo) of this species is locally known as frog vaccine which contains 16 bio-active peptides viz. adenoregulin, bombesin, bombesinnona peptide, a bradykinin derivate, caerulein, deltorphin, neurokinin B, phyllomedusin, phyllocaerulein, phyllokinin, phyllolitorin, preprotachykinin B, ranatachykinin A, sauvagine, T-kinin and urechistachykinin II . Unintentional intake may lead to nausea, vomiting, facial edema, palpitations and hypotension within few minutes (Hesselink, 2018; https://en.wikipedia.org/wiki/Phyllomedusa_bicolor)

Polypedates macrotis

Image credit: Rushenb, Wikipedia

Family: Rhacophorida

Common name: Dark-eared or masked tree frog

Global distribution: It is found distributed from Malaysia (especially southern Thailand) to Sumatra, Borneo and southern Philippines

Habitat: It is often found adjacent to small pools and puddles in flooded forest areas

Physical Description: It has a characteristic broad, dark stripe which is extending from behind the eye and along as about one-third of its flank. This stripe may break into separate patches. Its dorsal skin is mottled pale to medium brown, and it is pale ventrally. Lips are pale. Hind feet are extensively webbed, and fore feet are largely unwebbed. It grows up to 12 cm.

Harmful organs and harm: It is a toxic species as its skin has been reported to contain tetrodotoxin. In mouse test, the toxicity of this species had scores of 31-923MU/g (Tanu et al., 2001).

5.2.3. Newts

Cynops ensicauda popei (= Cynops ensicauda)

Image credit: Wikipedia

Order: Urodela Family: Salamandridae

Common name: Sword-tailed newt

Global distributionIt is restricted only to Okinawa and Anami Islands in Japan.

Habitat: It inhabits all kinds of water bodies such as ponds and streams as well as man made structures like rice paddies, road-side ditch

Physical Description:. It has a squarish head, bright orange belly and a clearly visible parotoid glands, In this subspecies, white, yellow or gold spots are present on the sides and tail and the background color is darker than its other subspecies. Sexual dimorphism is evident as the females are normally larger in size and are having a significantly longer tail and slender shape. Males have a tail which is as long as the body and is rounded at the end. It has a maximum total length of 18.0 cm for females and 13 cm for males. It is an endangered species.

Harmful organs and harm: When threatened, it produces a transparent skin toxin. It is poisonous when it is eaten or its secretion is ingested by mistake. Irritations may occur if open wounds are exposed to its secretions. It is also advised not to rub one's eyes or placing hands in mouth after this newt is handled (https:// okinawanaturephotography.com/)

The toxicity of this newt is due to its tetrodotoxin analogs viz. 6-epi TTX and 11-deoxy TTX extracted from the skin, egg, ovary, muscle and blood of this species (Yasumoto et al., 1988).

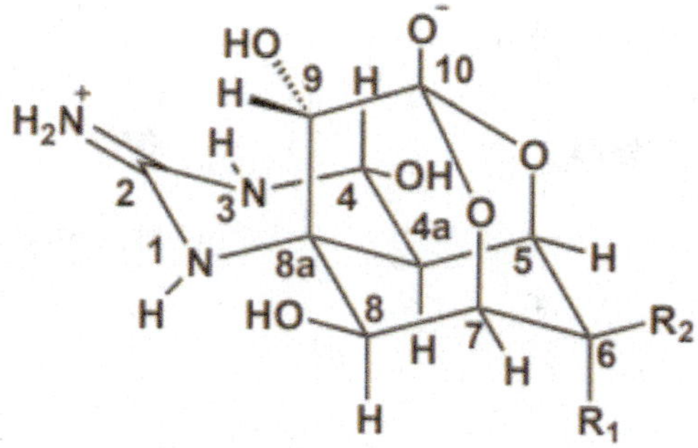

R1 R2

6-epi TTX 11-CH2OH OH

11-deoxy TTX OH 11-CH3

Tetrodotoxin analogs

Notophthalmus viridescens

Image credit: Patrick Coin (Patrick Coin) , Wikipedia

Order: Urodela Family: Salamandridae

Common name: Eastern newt

Global distribution: It is native to North America. It is found distributed throughout most of eastern North America

Habitat: It dwells commonly in small lakes, ponds, and streams

Physical Description: Colour of the adult newt is ranging from yellowish-brown to greenish-brown dorsally and it has black-bordered red spots. Its belly coloration is yellow with black spots. It has small eyes with a horizontal pupil. It grows to a maximum length of 13 cm.

Harmful organs and harm: Its skin,egg and ovary have been reported to contain tetrodotoxin and its analogue 6-epitetrodotoxin (Yotsu and Mebs,2001)

5.2.4. Salamanders

Salamandra salamandra

Image credit: Petar Milosevic, Wikipedia

Order: Urodela Family: Salamandridae

Common name: Fire salamander

Global distribution: Southern and central Europe

Habitat: It needs small brooks or ponds with clean water for the purpose of its breeding

Physical Description: It is black with yellow spots or stripes to a varying degree; some specimens can be nearly completely black while on others the yellow is dominant. Shades of red and orange may sometimes appear, either replacing or mixing with the yellow according to subspecies. Adults reach a size of 25 cm and weight of 40g.

Harmful organs and harm: The poison glands of this species are found concentrated in the head and the dorsal skin surface. The coloured regions of the animal's skin are usually found with more number of toxic glands. The toxic compounds of this species include samandarin , serotonin and tryptamine. Among these compounds, samandarin is a convulsive, more toxic than sodium cyanide. It has been reported to possess a powerful local anaesthetic effect and it acts on the central nervous system, leading to respiratory paralysis and death (Toledo and Jared,1995). Further, compounds in the skin secretions of this species may even be potentially dangerous to human life (https://en.wikipedia.org/wiki/Fire_salamander).

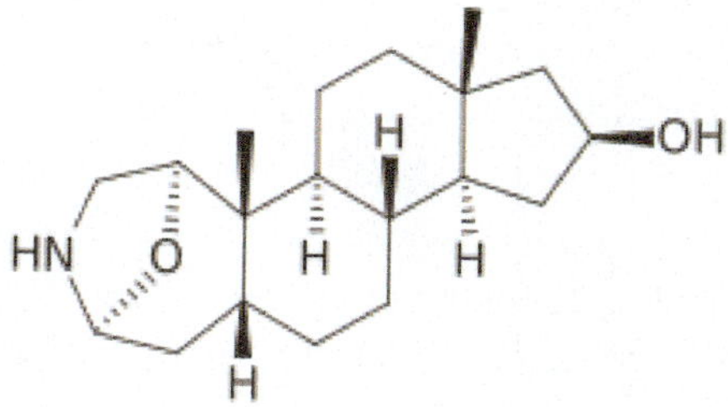

Samandarin

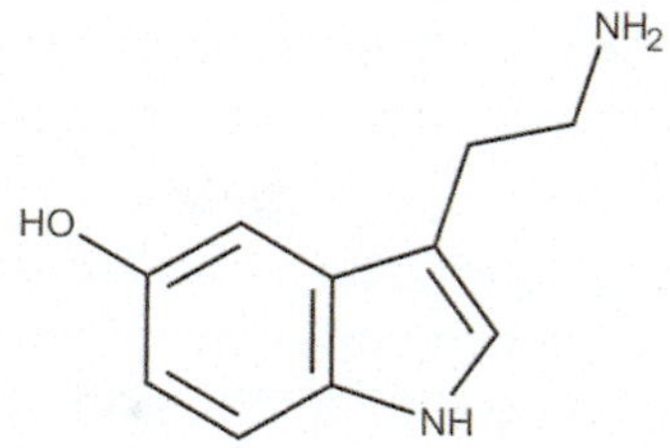

Serotonin

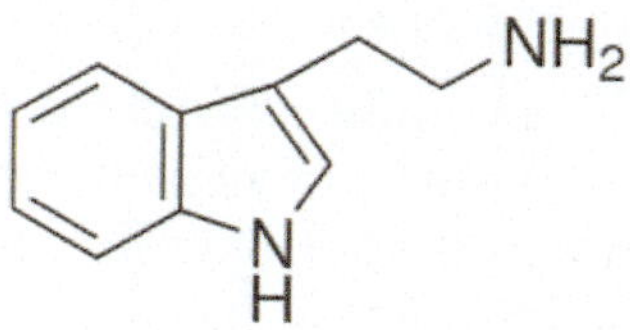

Tryptamine

Taricha granulosa

Image credit: Wikimedia Commons

Order: Urodela Family: Salamandridae

Common name: Rough-skinned newt or roughskin newt

Global distribution: North America: Santa Cruz, California, and Alaska

Habitat: It lives in all aquatic habitats such as lakes, ponds, ditches, and slow-moving streams

Physical Description: It is a stocky newt with rounded snout. Its dorsum coloration ranges from light brown to r brownish-black . Belly head, legs, and tail are from orange to yellow Skin is granular, but males are smooth-skinned especially during breeding season. It has smaller eyes and yellow irises, V-shaped teeth pattern and dark eyelids . Males can be distinguished from females during breeding season by their large swollen vent lobes and cornified toe pads. It reaches a maximum length of 18 cm.

Harmful organs and harm: The skin, egg, ovary, muscle and blood of this species contain the potent neurotoxin viz. tetrodotoxin (Mosher et al., 1965). Mailho-Fontana et al. (2019) reported that the skin of this species shows several evenly distributed poison glands (granular glands) which are composed of secretory cells. In addition to the presence of TTX in the poison glands, TTX was also identified within the mucous glands . The amount of TTX present in the skin of an individual newt of this species is sufficient to kill 25,000 mice (Mailho-Fontana et al., 2019).

Toxicity of this species is generally felt only if this newt is ingested, albeit some individuals have been reported to feel skin irritation after dermal contact, particularly if the eyes are touched after handling the animal without washing hands. Ingestion of this newt by mistake has also caused human fatality . The tetrodotoxin (TTX) (formerly called "tarichatoxin") of this species has been reported to target voltage gated sodium channels. The reverse binding to sodium channels in nerve cells blocked electrical signals necessary for conducting nerve impulses. This inhibition had the effect of inducing paralysis and death by asphyxiation (https:// en.wikipedia.org/wiki/Rough-skinned_newt)

Triturus cristatus

Image credit: Rainer Theuer, Wikipedia

Common name: Northern crested newt, great crested newt or warty newt

Global distribution: It is native to Great Britain; northern and central continental Europe and parts of Western Siberia

Habitat: It moves to aquatic breeding sites like fish-free ponds during spring.

Physical Description:It is a relatively large newt species having a maximum total length of 13.5 cm in males and 20.0 cm in females. Body shape of this species is normally related with its skeletal build. These newts have rough skin, and are dark brown on the back and sides. Black spots and heavy white stippling are seen on the flanks. Female has a yellow line running along the lower tail edge. Throat has a mixed yellow–black coloration with fine white stippling. Belly is yellow to orange with dark, irregular blotches.

Harmful organs and harm: The skin, egg, ovary, muscle and blood of this species possess toxic tetrodotoxin and its analogue 6-epitetrodotoxin (Yotsu- -Yamashita, et al., 2007). Toledo and Jared (1995) reported on the occurrence of biogenie amines in the venom of this species. A new protein of molecular weight 160,000 has been isolated from the skin venom of this species. This protein and its crude venom showed hemolytic effects of human erythrocytes *in vitro* in 0-24 h (Jaussi and Kunz, 1978)

5.3. Water Snakes

a) Cottonmouth

Among the water snakes, the cottonmouth bites are very dangerous animals. They are not poisonous, but they are rather venomous. Their venom may cause immense swelling, pain and tissue damage. Loss of arms and legs and even death have also been reported. Further, the bite of these snakes may also lead to extra infections owing to their eating of carrion and fangs which are accessed the bloodstram of affected persons.

Cottonmouth's fangs: The fangs of cottonmouth are hollow and twice the size of the rest of its teeth. When the fangs are in use, they are tucked against the roof of the mouth . Sometimes cottonmouths are known to shed their fangs and grow the new ones

Symptoms of cottonmouth bite: The symptoms of cottonmouth bite include nausea, pain. numbness, breathing trouble, vision problems and increased heart rate, As its venom is a hemotoxin, it has been reported to stop blood from coagulating by breaking down red blood cells. Consequently the circulatory system may begin to bleed out.

Treatment: As cottonmouth's bite often comes with a partial dose of venom, only wound care may be needed without antivenom. Its bite may not be fatal and it is better to seek medical treatment immediately if any body is bitten. Presently there are two types of cottonmouth antivenom in the USA. While one is derived from sheep, These antivenoms have been reported to boost the human immune response to the envenomation but they cannot reverse tissue damage, but they can stop it. (https://a-z-animals.com/blog/how-many-people-do-cottonmouths-water-moccasins-bite-per-year/)

b) Copperhead

The copperhead venom has been reported to be less potent than many other pit viper species. However the bite from copperhead snake is still considered as a serious health issue. Though hemotoxic, necrotizing, and anticoagulant effects are common with the copperhead venom, fatalities are rare in humans. Further, local tissue damage which is related to enzymatic proteins and coagulopathy (blood clotting issue) has been found to be rare with copperhead bites. Common symptoms associated with the copperhead bites are hyperthermia, local swelling, pain, redness, and bruising (https://www.petpoisonhelpline.com/blog/copperhead-snakes-envenomation-risk-for-animals/)

C) Rattlesnakes

The rattlesnakes which inhabit normally the wettest swamps are dangerous but their bites are very rarely fatal. However, if the bites of these snakes are left untreated, they may result in severe medical problems or may even be fatal. The venom of the rattlesnake bites containing hemotoxic elements.has been reported to damage tissue and affect the circulatory system by destroying skin tissues and blood cells and by causing internal hemorrhage.

Symptoms of rattlesnake bite: The common symptoms of rattlesnake bite include: numbness in the face or limbs, vomiting, salivating, blurred vision and breathing difficulty.

Treatments: It is better to seek medical assistance immediately after the bite. In order to minimize the risk and before admission in the emergency department, the following guidelines may be followed (https://www.healthline.com/health/rattlesnake-bite#symptoms)

1. Affected area should not be raised above the level of the heart.
2. No unwanted movements
3. Tight clothing or jewelry should be removed
4. Wound may be allowed to bleed
5. Wound should not be washed
6. A clean bandage may be placed on the wound
7. If signs of shock are felt, feet may be raised slightly
8. Tourniquet should not used and water or ice should not be applied on wound

Hydrophis semperi

Image credit: . Angel C. Alcala (Reproduced with permission)

Class: Reptilia Order: Squamata Family: Elapidae

Common name: Lake Taal snake, Philippine freshwater sea snake, Luzon dovetail

Global distribution: It is found only in a single lake on the island of Luzon in the Philippines

Habitat: It is found almost-exclusively in freshwater. This sea snake has adapted to live in fresh water.

Description: Like all "true" sea snakes, it is elongated with a small head. It has the characteristic flattened, paddle-like tail. Adult specimens have a total maximum length (including tail) of 70 cm. It has a dark blue or black body with narrow yellow or white crossbands. Its nostrils are located almost dorsally, with valves which prevent water from entering its nasal and oral cavities when the snake is submerged.

Harmful organs and Harm: Watt and Theakston (1985) reported that this freshwater venomous seasnake caused eight definite bites and one probably attributable death in a a freshwater lake of Philippines. In contrast to previous seasnake studies, there was an absence of asymptomatic cases.

Laticauda crockeri

Image credit: California Academy of Sciences (CC)

Order: Squamata Family: Elapidae

Common name: Crocker's sea snake

Global distribution: It is endemic to a single brackish lake (Lake Te-Nggano) on Rennell Island in the Solomon Islands; Taiwan and Solomon

Habitat: This demersal species lives both in freshwater and brackish-water (https://reptile-database.reptarium.cz/species?genus=Laticauda&species=crockeri)

Description: It is a dwarf species with black or very dark brown pigmentation on its body. Nasals are separated by internasals . It has a strongly seasonal pattern of reproduction..

Harmful organs and Harm: It is a venomous snake (https://reptile-database.reptarium.cz/species?genus=Laticauda&species=crockeri) with the presence of a short neurotoxin B (https://www.uniprot.org/uniprot/P25496).

Naja annulata (= *Boulengerina stormsi*)

Image credit: Wikipedia

Order: Squamata Family Elapidae

Common name: Ringed water cobra or banded water cobra

Global distribution: It is native to western and central Africa

Ecology: It lives near and in rivers, streams and lakes.

Description/ Biology: This species is one of the two species of water cobra in the world (other one is *Naja christyi*) . It is a large, heavy-bodied snake with a short, broad and flat head. It has medium-sized dark eyes with round pupils. Body is cylindrical and tail is long. Scales are smooth and glossy and are in 21-23 rows at its midbody. Adults snakes grow to a maximum length of 2.8 m. Scales are smooth. It is capable of spreading a narrow hood. Body coloration is mostly grayish-brown or reddish-brown with black bands all along the body. Belly is pale yellow, while the tail is completely black.

Harmful organs and harm: It is a venomous species (https://reptile-database.reptarium.cz/species?genus=Naja&species=annulata). The LD50 of its long-chain neurotoxins from the venom of this species was 0.086 mg/kg (https://en.wikipedia.org/wiki/Naja_annulata)

Naja christyi (= *Boulengerina christyi*)

***Boulengerina* sp.**
Image credit: Artist: P. J. Smit, Wikipedia

Family: Elapidae

Common name: Congo water cobra or Christy's water cobra

Global distribution: It is native to Sub-Saharan Africa

Ecology: It is found most commonly in or near banks of lakes, rivers, and streams. It is limited to the lower Congo River

Description/ Biology: It is a moderately long and slender-bodied snake with a tapering tail. Head is short and is distinct from the neck. Neck is capable of spreading into a hood. Eyes are small in size with rounded pupils. Dorsal scales are smooth and shiny. Body coloration is blackish browndorsally. Nape and the anterior portion of the body have irregular pale brown and black cross bars. Upper lip is pale brown. Lower surface of the head is brownish white. Belly and ventral surface of the tail are blackish. It grows to maximum length of 2.3 m.

Harmful organs and harm: It is a venomous species and it is believed that its venom is dangerously neurotoxic, like that of most of the elapids. The LD_{50} of long-chain neurotoxins from the venom of this species was found to be 0.090 mg/ kg (https://en.wikipedia.org/wiki/Naja_christyi)

Agkistrodon contortrix

Image credit: Wikimedia Commons

Family: Viperidae

Common name: Eastern copperhead, red snake or white oak snake

Global distribution: North America

Ecology: It also occurs in riparian habitats; near permanent or semipermanent water ; and low-lying, swampy regions.

Description/ Biology: Adults of this species grow to a maximum size of about 1m. Males are generally larger than females. Body is stout and head is broad and distinct from the neck. Coloration of body consists of a pale tan to pinkish-tan ground color which becomes darker towards the foreline. A series of 10–18 crossbands are seen on the body. Coloration of belly is the same as that of ground color, but partly it may be a little. At the base of the tail , there are 1-3 brown crossbands which are followed by a gray area.

Harmful organs and harm: Its venom has been reported to possess a protein called contortrostatin. Although these snakes are venomous, they are generally not aggressive and bites are seldom fatal. Its venom has a lethal dose around 100 mg, Bite symptoms include severe pain, tingling, swelling, and acute nausea (https://en.wikipedia.org/wiki/Eastern_copperhead).

Agkistrodon piscivorus

Image credit: Scott Zona, Wikipedia

Common name: Cottonmouth or water moccasin

Global distribution: It is native to the southeastern United States.

Ecology: It dwells in creeks, streams, marshes, swamps, ponds, river bottoms and lakes; and it is semiaquatic and prefers moist habitats in close proximity to water..

Description/ Biology: It is the only world's semiaquatic viper. Females are generally smaller than males which reach a maximum length of even 188cm. Its broad head is distinct from the neck, and snout is blunt in profile. Its body color pattern may consist of a brown, yellowish-olive, or blackish which is overlaid by a series of 10–17 dark brown to black crossbands. Head is brown colored. While eastern snakes have a broad, dark, postocular stripe, bordered with pale pigment above and below, it is absent in western populations. Underside of the head is whitish or cream.

Harmful organs and harm: This species is capable of delivering a fatal bite. Its venom is cytotoxic and proteolytic and it may lead to necrosis. Although deaths due to the biting of this snake are rare, it may leave scars, and on certain occasions amputation may result. .Symptoms of bite include ecchymosis (discoloration of the skin) and swelling (https://animaldiversity.org/accounts/Agkistrodon_piscivorus/).

Thackston and Wills (2022) reported that the venom of this species contains phospholipase A2 (PLA2), an enzyme that hydrolyzes phospholipids. Further, the presence of metalloproteinases (including fibrolase, a fibrinolytic enzyme) in the venom of this species have been reported to contributes to hematologic and soft-tissue toxicity. Hematologic effects activate the coagulation cascade as well as have fibrinolytic, prothrombin activation and inhibit platelet aggregation. On the other hand, the soft-tissue toxicity can cause myonecrosis, skin injury, and robust inflammatory response

Crotalus horridus

Image credit: Wikipedia

Family: Viperidae

Common name: Venomous pit viper or timber rattlesnake

Global distribution: North America; Canada and USA

Ecology: It occurs in areas around swamps and river floodplains (Wikipedia)

Description/ Biology: This species is characterized by the presence of a rattle (it is composed of horny, loosely connected hollow segments, one of which is added every time the snake sheds its skin) and a short tail; absence of a nasal pore and position of its facial pit which is below the naso-ocular line.

Harmful organs and harm: It is venomous and it can harm humans if they are bitten (https://animaldiversity.org/accounts/Crotalus_horridus/; https://reptile-database.reptarium.cz/species?genus=Crotalus&species=horridus). Four venom patterns have been identified in this species and they are neurotoxic A type; hemorrhagic and proteolytic B type; a weak C type venom; and D type venom with a small basic peptide (which works as a myotoxin), a fibrinogen-clotting enzyme and bradykinin-releasing enzyme. Among these venom types, type A contributes significantly to the overall toxicity of this species (https://en.wikipedia.org/wiki/Timber_rattlesnake#Distribution_and_habitat)

Sistrurus miliarius

Image credit: Wikipedia

Family: Viperidae

Common name: Pygmy rattlesnake or Carolina Pigmy Rattlesnake

Global distribution: It is endemic to the Southeastern United States.

Ecology: It occurs in wet prairies, swamps, floodplains ,moist lowlands, floodplains, near lakes and marshes.[

Description/ Biology: It is a short, and thick-bodied snake. It has a dark line through the eye on each side of its face. A series of dark, roughly circular spots are seen running down the center of the back. These spots interrupt a thin reddish-orange stripe which runs along the midbody line. A row of dark blotches is seen along each side of the body. Background coloration of the body is generally a light gray or brown, but specimens with pinkish or reddish background coloration have also been recorded. It grows to a maximum length of 80 cm.

Harmful organs and harm: These snakes have been reported to bite humans when they are threatened. Venom injected by the bite of this species is fairly toxic. No human deaths due to the bite of this species have been reported and it is largely due to its small amounts of venom delivery. . Symptoms associated with the bite include pain, swelling, discoloration, blisters, and vomiting. Envenomation of this snake can also cause blood cell loss and a reduction in blood circulation. If left untreated, bites can result permanent tissue damage. Bites involving children have resulted necrosis and prolonged hospitalization (https://animaldiversity.org/accounts/Sistrurus_miliarius/).

5.4. Alligators and Crocodiles

Alligator mississippiensis

Image credit: TimVickers, Wikipedia

Order: Crocodilia Family: Alligatoridae

Common name: American Alligator

Global distribution: It is native to the Southeastern United States

Ecology: It lives largely in marshes, swamps, rivers, ponds, and lagoons. It is also seen in man-made water structures such as swimming pools and fishponds areas

Description : Snout of this species is round. Upper jaw is larger than the lower jaw. It is a homodont and has about 80 teeth of uniform size. As its teeth often breakor fall out, they are replaced immediately throughout its lifetime. Coloration of its body is olive black . It has been reported to attain a length of 44 cm and weight of 408 kg.

Harmful organs and harm: These alligators have been reported to cause harm or death to humans. Between 1948 and 2005, 337 attacks occurred in Florida, 15 in Texas, 9 in Georgia, 9 in South Carolina, 5 in Alabama, 2 in Louisiana, and 1 in North Carolina. Further in Florida, between 1948 and 2012, 22 of the 344 attacks were fatal. (Wikipedia). Alligators that are more than 2.7 m have been reported for the most attacks. Attacks can happen throughout the year, and are known to generally occur when people are swimming (https://animaldiversity.org/accounts/Alligator_mississippiensis/).

Alligator sinensis

Image credit: Stolz, Gary M., Wikipedia

Common name: Chinese alligator

Global distribution: China's Yangtze River region

Ecology: It occurs in slow-moving freshwater rivers , streams, lakes, ponds and swamp. It is also known to prefer remnant wetlands and ponds surrounded by agricultural lands.

Description/ Biology: It is the smallest species of crocodilians growing to only about 2 m in length and it attains a maximum weight of 45 kg. It is black or dark gray in color. Head is robust with a short and broad snout which points slightly upwards and narrows at the end. It is fully armored including its belly. There are 17 rows of scales across its body and scales of belly and side are soft .

Harmful organs and harm: It is potentially dangerous and if provoked it would attack (https://animaldiversity.org/accounts/Alligator_sinensis/)

Melanosuchus niger

Image credit: Wikipedia

Order: Crocodilia Family: Alligatoridae

Common name: Black Caiman

Global distribution: It is found distributed throughout the Amazon Basin; its range includes most of northern and central South America

Ecology: It is commonly found in slow-moving freshwater rivers, lakes, wetlands, swamps, and seasonally flooded areas

Description/ Biology: It is a long, squat animal with big jaws, long tail and short legs. It has thick, scaled skin, and its eyes and nose are located on the top of its head. Body is almost black in color. Head is large and heavy. Its lower jaw has grey to brown banding. It grows to a maximum length of 6m.

Harmful organs and harm: Several attacks of this species on humans have been reported (https://animaldiversity.org/accounts/Melanosuchus_niger/). In Amazon region a 11-year-old child was fatally attacked by this species (Haddad Jr and Fonseca,2011).

Crocodylus acutus

Image credit: Tomás Castelazo , Wikipedia

Family: Crocodylidae

Common name: American Crocodile

Global distribution: Americas: from South Florida and Mexico to Peru and Venezuela.

Ecology: It inhabits freshwater bodies such as rivers, lakes and reservoirs; and it can also be found in brackish environments, such as estuaries and swamps.

Description/ Biology: It is one of the larger crocodile species. Males are much bigger than females in size and weight. Males can attain a maximum length of 6.1 m and a weight of 907 kg . It has about 40 teeth in its upper jaw and 30 teeth in the lower jaw. Teeth are conical, and sharply pointed; and they may vary in size all along the length of the jaws. Its overall coloration ranges from gray, to greenish gray with dusky markings.

Harmful organs and harm: These crocodiles have been reported to be potentially dangerous to humans and attack and kill the victims on rare occasions (https://animaldiversity.org/accounts/Crocodylus_acutus/).

Crocodylus palustris

Image credit: Hadi Karimi , Wikipedia

Common name: Mugger Crocodile , marsh crocodile, broad-snouted crocodile or mugger

Global distribution: It is found distributed from southern Iran to the Indian subcontinent

Ecology: It dwells in freshwater lakes, hill streams, village ponds, rivers, marshes and slow-moving, shallow water bodies. It also thrives in artificial reservoirs and irrigation canals.

Description/ Biology: Adult mugger crocodiles are dark olive to grey or brown. Head is rough without any ridges. Large scutes are seen around the neck which is well separated from the back. Limbs have keeled scales with serrated fringes on outer edges. Outer toes are well webbed. Snout is slightly longer than broad with 19 upper teeth on each side. Symphysis of lower jaw extends to the level of fifth tooth.

Harmful organs and harm: These crocodiles are potentially dangerous and they have been reported to attack fishermen- (https://animaldiversity.org/accounts/Crocodylus_palustris/). In the 1990s, several fatal attacks of muggers on humans have been reported in India (https://en.wikipedia.org/wiki/Mugger_crocodile)

Crocodylus niloticus

Image credit: Dewet , Wikipedia

Common name: Nile Crocodile

Global distribution: It is found distributed throughout sub-Saharan Africa, Nile Basin, and Madagascar

Ecology: It is common in rivers, freshwater marshes, and mangrove swamps. It is rarely found in saltwater and brackish lakes.

Description/ Biology: It is the second-largest living reptile in the world, after the saltwater crocodile (*Crocodylus porosus*). Males have been reported to attain a maximum length of 6.1 m and weight of 1089 kg. It has a thick, scaly, strongly armoured skin. Dorsum of adult Nile crocodiles is with a dark bronze colouration and faded blackish spots and stripes which are variably appearing across the back. Belly is dull yellow. Flanks are yellowish-green in colour and possess dark patches which are arranged in oblique stripes in much variable patterns.

Harmful organs and harm: It is one of the most dangerous species of crocodile owing to its extremely powerful bite and this species has been reported to be responsible for hundreds of human deaths every year (https://en.wikipedia.org/wiki/Nile_crocodile)

5.5. Mammals

Hippopotamus amphibious

Image credit: Diego Delso , Wikipedia

Class: Mammalia Order: Artiodactyla Family: Hippopotamidae

Common name: Hippopotamus

Global distribution: It is found exclusively in the Ethiopian region of the world; throughout the savanna of Africa and the main rivers of Central Africa.

Ecology: This semi-aquatic mammal inhabits shallow lakes, rivers, and swamps

Description/ Biology: This species attains a maximum length of 5m; weight of 3.2 tons and height of 1.7 m. Body which is bulky and barrel-shaped is purple gray or slate colored. Brownish pink coloring is seen around its eyes and ears. While its body is covered by scarce amount of thin hair, head and tail are with thick bristle like hair. Jaws are capable of opening up to 150 degrees and are with sharp, incisors and canine tusks. While its canines grow to 50 cm, incisors grow to 40 cm. While males reach a maximum length of 505 cm ,females only reach 345 cm. Maximum recorded weight of this species is about 4.5 tons.

Harmful organs and harm: This species is considered most dangerous and highly aggressive. Its bite force measures 12600 kPa (kilopascals) compared to that of lion (4500kPa) (Haddara e al., 2020). When these animals are threatened, especially in drought-burdened seasons, they have been reported to attack humans. Accidental encounters with human fatality have been recorded with hunters and boaters (https://animaldiversity.org/accounts/Hippopotamus_amphibius/)

CHAPTER 6

HARMFUL MICROBES OF FRESHWATER BODIES

6.1. Pathogenic Bacteria of Freshwaters and their Infections

Terrestrial freshwater systems which include groundwater, lakes, rivers, streams and wetlands contribute to private and municipal sources of drinking water. All these water bodies afford opportunities for exposure to infectious pathogenic microorganisms such as bacteria and fungi through recreational and occupational activities and wastewater discharges and contamination with human or animal feces. Waterborne diseases due to bacterial pathogens have also been reported to affect both developing and developed countries. In the USA alone , a total of 560,000 people suffer from severe waterborne diseases each year , and 7.1 million suffer from a mild to moderate infections, resulting in an estimated 12,000 deaths a year (Cabral,2010). The pathogenic bacterial species recorded worldwide from different freshwater bodies are given below.

Ground water

- *Bacillus megaterium*
- *Enterococcus* spp.
- *Escherichia coli,*
- *Pseudomonas aeruginosa*
- *Salmonella* spp.
- *Staphylococcus aureus*

- *Staphylococcus* spp.
- *Streptococcus faecalis*

Rivers

- *Campylobacter jejuni*
- *Escherichia coli*
- *Pseudomonas aeruginosa*
- *Salmonella* spp.
- *Yersinia enterocolitica*

Lakes and reservoirs

- *Cryptosporidium*
- *Cryptosporidium oocysts*
- *Escherichia coli*
- *Serratia marcescens*
- Source: Pandey et al., 2014

Enteric and opportunistic pathogenic bacterial species

- Ashbolt (2015) reported on the enteric and water-based bacterial species as detailed below

Enteric (waterborne) bacteria

- *Acinetobacter baumannii*
- *Arcobacter butzleri*
- *Helicobacter pylori*
- *Clostridium difficile*
- *Listeria monocytogenes*
- *Pseudomonas aeruginosa*
- *Staphylococcus aureus*
- *Yersinia enterocolitica*
- *Aeromonas hydrophila*
- *Campylobacter coli*
- *Campylobacter jejuni*
- *Salmonella enterica (non-typhoid)*

- *Shigella sonnei*
- *Vibrio cholerae*

Water-based (opportunistic) bacteria

- *Acinetobacter baumannii*
- *Aeromonas hydrophila*
- *Afipia* spp
- *Bosea*, spp
- *Coxiella burnetii*
- *E. coli (toxigenic strains)*
- *Listeria monocytogenes*
- *Staphylococcus aureus*
- *Stenotrophomonas maltophilia*
- *Legionella longbeacheae*
- *Legionella micdadei*
- *Legionella pneumophila*
- *Escherichia coli O157:H7*
- *Pseudomonas aeruginosa*
- *Parachlamydia* spp.

Water-Based (Environmental) bacteria

- *Legionella pneumophila*
- *Legionella longbeacheae*
- *Legionella micdadei*
- *Mycobacterium avium*
- *Mycobacterium intracellulare*
- *Mycobacterium chelonae*
- *Mycobacterium fortuitum*
- *Mycobacterium gordonae*
- *Mycobacterium kansasii*

Source: Ashbolt, 2015

Bacterial diseases transmitted through water

The most important bacterial diseases which are transmitted through water on humans include cholera, gastroenteritis, typhoid fever , salmonellosis and shigellosis. The causative bacterial species for these diseases are given below

- **Most important bacterial diseases transmitted through water**

Disease	Causal bacterial agent
Cholera	*Vibrio cholerae*
Gastroenteritis	*Vibrio parahaemolyticus*
Typhoid fever and salmonellosis	*Salmonella enterica*
Bacillary dysentery (shigellosis)	*Shigella dysenteriae* *Shigella flexneri* *Shigella boydii* *Shigella sonnei*
Diarrhea and gastroenteritis	*Escherichia coli*

Source: Cabral,2010

Vibrio cholerae and cholera

This species is largely responsible for cholera (often called Asiatic cholera or epidemic cholera), a severe diarrheal disease in humans. This pathogen is invariably transmitted to humans by water or food. It is a noninvasive species and is affecting the small intestine of humans through its secretion of an enterotoxin called cholera toxin which acts on the mucosal epithelium. In its extreme manifestation, this cholera may result in fatal illness of the patient. It has been reported that a healthy person may become hypotensive within one hour of the onset of symptoms and he may die within 2-3 hours if no proper treatment is provided (Pandey et al., 2014)

Vibrios, the causative agents of cholera are the most common organisms in surface waters of the world. They occur in both marine and freshwater habitats. Waterborne species of *Vibrio such* as *Vibrio alginolyticus, V. cholerae , V. fluvialis, V. furnissii, V. harveyi, Grimontia hollisae (= Vibrio hollisae), V. mimicus, V. parahaemolyticus,* and *V. vulnificus* have been reported to be the causative agents for the cholera infection in humans. Among these species, the most important species for this infection is *V. cholerae* (Cabral,2010). The flesh-eating bacteria, *V. vulnificus* which is common in freshwater systems has been reported to cause infections in people who were swimming, crabbing, and boating in the Long Island Sound. Available reports say that five people in the age group 49 and 85 years were severely infected by this bacterial species. Among them, two showed septicemia, an infection in the bloodstream, and three displayed serious wounds (https://www.health.com/condition/infectious-diseases/flesh-eating-bacteria-long-island-sound).

Salmonella and salmonellosis

Salmonellosis is a symptomatic infection caused by the bacteria of the Salmonella type. The most common symptoms associated with this infection include diarrhea, fever, abdominal cramps, and vomiting. Among the subspecies of

Salmonella enterica, S. enterica subsp. *enterica* and its serovars (variations) are largely responsible for human pathogenicity. Further, the serovar Enteritidis is the most frequently isolated one from humans all over the world. The different serovars of *S. enterica* subsp. *enterica* and their infections in humans are given in the following table.

Species	Sub-species	Serovar (variation)	Pathogenicity in humans
Salmonella enterica	*S.enterica* subsp. *enterica*	Choleraesuis	Pathogenic to humans
-	-	Enteritidis	Gastroenteritis in humans
-	-	Paratyphi A-C	Paratyphoid in humans
-	-	Typhi	Typhoid in humans

Source: Cabral,2010; Pandey et al., 2014

Shigella and Shigellosis or Bacillary dysentery

The species of *Shigella* are not uniformly distributed in the world. The common disease causing species viz. *Shigella dysenteriae* is commonly found in densely populated regions of South America, Africa and Asia. The intestinal infection caused by the bacteria of *Shigella* spp. is called shigellosis. These pathogenic species may also be passed by drinking unsafe water or by swimming in contaminated water bodies. Children under age 5 are most likely to be affected by shigellosis, but it can also occur at any age. The symptoms associated with this infection include diarrhea (often with blood or mucus), stomach pain or cramps, fever and nausea or vomiting.

Shigellosis infection scenario

The total number of Shigella cases which occur each year throughout the World has been estimated at 165 million, including 163 million cases in developing countries. While a total of 1.1 million succumbed to death and 61% of all deaths involved children under age five . Important epidemics due to shigellosis have been reported in the last decades: (i) in 1970 , 112,000 people were affected and 13,000 died in Central America; (ii) in 1985, 5,000 people were found infected after ingestion of contaminated lettuce in Texas (USA); (iii) domestic cases of *Shigella sonnei* infection were recorded in several European countries, such as Norway, Sweden, and the United Kingdom in 1994; and (4) in 1996, 153 patients with shigellosis were recorded in Paris (Cabral,2010).

Escherichia coli strains

Escherichia coli strains isolated from the intestinal diseases have been grouped into at least six different main groups, based on the clinical features of the disease. Of these groups, ,the serotypes viz. enterotoxigenic *Escherichia coli* , enterohemorrhagic *Escherichia coli* and enteroinvasive *E. coli* are of more importance and are normally transmitted through contaminated water. Among these three serotypes, enterotoxigenic *E. coli* (ETEC) has been reported to cause infantile gastroenteritis. Ingestion of ETEC contaminated water may cause profuse watery diarrhea leading to dehydration and malnutrition especially in young children (Cabral,2010).

Emerging Waterborne Bacterial Pathogens

Mycobacterium avium complex (Mac)

The *Mycobacterium avium* complex (Mac) is consisting of 28 serovars (or serotype) of two different species viz. *Mycobacterium avium* and *M. intracellulare*.. Mac organisms are found in a wide range of aquatic habitats viz. marine waters, rivers, lakes, streams, ponds, etc. Mac organisms have been isolated from natural water and drinking water distribution systems in several developed countries including USA. The members of this complex are considered opportunistic human pathogens which commonly infect the immunocompromised people, (particularly people with HIV and AIDS) ingesting potable water containing these pathogenic organisms. The symptoms of the Mac infections often result from the colonization of either the respiratory or the gastrointestinal tract, with possible spreading to other locations in the body (Cabral,2010).

Helicobacter pylori

Helicobacter pylori has been reported as a major causative agent for gastritis and it has been implicated in the pathogenesis of peptic and duodenal ulcer disease and gastric carcinoma. Though the contaminated water and food seem to be of lesser direct importance, they can still play a significant role in association with improper sanitation and hygiene (Cabral,2010).

Aeromonas hydrophyla

Aeromonad pathogens , such as *Aeromonas hydrophyla, A. caviae* and *A.sobria,* have been isolated from contaminated water sources. Among these species *A. hydrophyla* is a dominant and highly infectious species and is often isolated from aquatic environments such as clean rivers and lakes. This species has been implicated as a potential causative agent of gastroenteritis, septicemia, meningitis, and wound infections. It has also been reported to play a significant role in the intestinal disorders in children, the elderly, and immunosuppressed people (Cabral,2010).

6.2. Pathogenic Fungi of Freshwaters and their infections

Pathogenic freshwater fungi

The private and municipal sources of drinking water, groundwater, lakes, rivers, streams and wetland are often associated with several species of waterborne (enteric) and water-based (opportunistic) pathogenic fungi due to runoff or sewage contamination. While the lakes, rivers, streams and wetland afford opportunities for exposure to these infectious fungi through recreational and occupational activities, both tap water and municipal water infect humans when they are ingested. Pathogenic fungi may also be associated with or be harbored by free-living amoebas in different freshwater bodies. Ashbolt (2015) and Baumgardner (2017) reported the following species of fungi are largely pathogenic and are the causative agents for several diseases.

- *Acremonium* spp.
- *Aspergillus fumigatus*
- *Aspergillus terreus*
- *Candida albicans*
- *Candida parapsilosis*
- *Candida albicans*
- *Cladosporium* spp.
- *Encephalitozoon intestinalis*
- *Encephalitozoon bieneusi*
- *Exophiala dermatitidis*
- *Fusarium* spp.
- *Paecilomyces* spp.
- *Penicillium* spp.
- *Trichoderma* spp.

Infections of pathogenic freshwater fungi

Eye Infections due to fungi from Freshwater

Fungal keratitis : symptoms- present with pain, redness, decreased visual acuity, photophobia, and increased tearing and discharge from the affected eye.25 Eyelids also may be infected and demonstrate blepharospasm, and the cornea may be hazy or have an area of opacification

The filamentous fungi such as the species of *Aspergillus* and *Fusarium keratoplasticum* ; the dematiaceous (dark-pigmented) fungi , *Scedosporium apiospermum* and *Candida* spp. have been reported to be the primary causes of fungal keratitis in most parts of the world. This disease is also much more common in developing countries. In the United States, *Candida* species were the predominant causes of fungal keratitis in a New York series (Baumgardner, 2017)

Microsporidial keratoconjunctivitis : Microsporidia, a diverse group of several genera have been reported to cause the fungal keratoconjunctivitis (microsporidial keratoconjunctivitis) in immuno compromised and normal hosts (often in contact lens users). A large outbreak of this disease occurred following heavy rain (and a potential association with muddy water exposure) during a rugby tournament in Singapore (Baumgardner, 2017).

Conjunctivitis with episcleritis and anterior uveitis : An outbreak of the acute conjunctivitis with episcleritis and anterior uveitis was reported in in Brazil due to freshwater exposure to *Emmonsia* sp. (adiaspiromycosis) (Baumgardner, 2017).

Exogenous endophthalmitis: The species of *Fusarium* and *Candida* are believed to be responsible to this disease. Risks of this infection include neutropenia (person with a low level of neutrophils) , human immunodeficiency virus/AIDS and intravenous drug abuse. Symptoms associated with this disease are pain, redness and blurred vision (Baumgardner, 2017).

Acute otitis externa: The species of fungi associated with this infection are of *Candida* and *Aspergillus*. Symptoms of this disease are hearing loss and lack of fetid discharge. Signs of this infection include ear canal erythema and edema. Other factors for candidal otitis externa are the swimming in public pools/baths and diabetes mellitus (Baumgardner, 2017)

Other infections : Other reported infections due to freshwater pathogenic fungi include central nervous system infections (including meningitis and brain abscess) , deep soft tissue or bone infections, and pulmonary infections, which are generally occurring in immunocompetent patients. Causative agents of these infections are the species of *Cladophialophora, Fusarium, Aspergillus, Pseudallescheria* and *Scedosporium*. Infection of the respiratory tract, systemic invasive disease (usually in immunocompromised hosts) and keratitis have also been reported with the pathogenic species of fungus viz. *Scedosporium apiospermum* (Baumgardner, 2017).

Fungal toxicity to humans: Humans may also be directly exposed to fungal toxins through the ingestion of contaminated water. The harmful aquatic fungal species *Aspergillus fumigatus* was isolated from patients suffering from immunosuppression infections and organ failure (Magwaza et al., 2017).

CHAPTER 7

HUMAN HEALTH HAZARDS ASSOCIATED WITH HARMFUL FRESHWATER LIFE: CASE REPORTS

In the freshwater biome , all types of water pollution have already been reported to be harmful to the health of humans and animals especially after long term exposure. Recent findings have however shown that certain inhabitants of this biome particularly freshwater harmful algal blooms (FHABs) and venomous fishes are very hazardous to humans owing to their toxins or stings. This chapter deals with the case reports relating to human health hazards associated with certain harmful freshwater algae and animals .

7.1. Harmful Algal Blooms

The incidence of freshwater harmful algal blooms (FHABs) has increased considerably especially in the USA especially during the last three decades .During the period 2009 – 2010, 11 FHAB-associated outbreaks have been reported in this country. Case reports relating to a wide range of acute illnesses among human beings suggested that these outbreaks were largely due to the recreational exposure to cyanobacteria (blue-green algae) and their toxins (Trevino-Garrison et al., 2015). Anon. (2019) reported that during Jan-Dec. 2019, 14 states of USA had 181 FHAB events in lakes/reservoirs (75%), rivers/streams (15%) and ponds (8%). All these events happened during summer/fall (June–October). It was also found that the toxin, microcystin was the most common one in about 130 events and the cyanobacterial genus *Microcystis* showed its presence in about 30 events (Trevino-Garrison et al., 2015).

Case report 1

Locality: Merced River near Hites Cove ,California, USA

Description: A family of 3 people including a one-year-old female child and a dog died and this was reported to be due to the toxic algae blooms associated with high concentration of algal cells in the Merced River near Hites Cove ,California. The details of death of this family were however, not known. Following this incident, the Sierra National Forest (SNF) informed the visitors who liked to enjoy the Merced River and SNF, not to wade , swim, or permit their pet animals to enjoy this water resource as some species of algal mats of this rive may produce toxins, which may pose health problems to humans and pet animals (O'Neil , 2021).

Case report 2.

Locality: Toledo, Ohio,USA

Description: During 2014, the formation of a harmful cyanobacterial bloom in Lake Erie affected the drinking water for more than 500,000 people (https://crsreports.congress.gov/product/pdf/R/R44871,2020).

Case report 3

Locality: Florida ,USA

Description: The formation of a massive harmful algal bloom in the Florida's Lake Okeechobee sffected aquatic life and tourism heavily (https://crsreports.congress.gov/product/pdf/R/R44871,2020).

Case report 4

Locality: Milford Lake, Kansas, USA

Description: The FHAB which appeared in the Milford Lake, Kansas, USA during 2011 was found to be responsible for human illness. A total of 7 people in the age of 17-63 were found affected and hospitalized. They possessed the primary symptoms such as eye and upper respiratory tract irritation, rashes and gastrointestinal problems which were due to their direct contact , ingestion or inhalation of the bloom (Trevino-Garrison et al., 2015)

Case report 5

Locality: Milford Lake, Kansas, USA

Description: A 17-year-old male was admitted to the emergency department on July 23, 2011. He had symptoms such as cough, headache, fever and sore throat. As per his statement, he developed these symptoms about 24 hr after swimming in the Milford Lake, Kansas, USA. This case report was found associated with the formation of a potential FHAB in this lake on July 18, 2011. This bloom showed the presence of cyanobacteria cells and their toxin, microcystin in high concentrations (Trevino-Garrison et al., 2015)

Case report 6

Locality: Milford Lake, Kansas, USA

Description: A 38-year-old male was admitted to the emergency department on August 8, 2011 due to cyanobacteria toxicosis. He had symptoms such as joint pain, headache, sore throat and fever. He reported swallowing water as he fell in the lake while he was water skiing at the Milford Lake on August 5, 2011. Due to high cyanobacteria cell concentrations and microcystin toxin levels, this lake was under a public health warning on the date of illness of this patient (Trevino-Garrison et al., 2015).

7.2. Poisonous Plants

Case report 1: Water hemlock poisoning

A 23-year-old male was admitted to the emergency department as he experienced multiple tonic-clonic seizures which was however terminated with the administration of midazolam (injection is used to produce sleepiness or drowsiness). The unintentional drinking of the tea made from the roots of the poisonous plant *Cicuta maculata* was finally found to be the reason for his illness. This was ascertained after several diagnostic procedures and first aid treatments (Butterfield et al., 2019)

Case report 2: Hemlock water-dropwort poisoning

This case report deals with two persons who happened to ingest the roots of the poisonous plant *Oenantha crocata* (hemlock water-dropwort and is sometimes called as dead man's fingers) by mistaking them for edible tuberous radish. Among these two patients, one showed vomiting and was given one day symptomatic treatment. On the other hand, the other person had severe problems including prolonged convulsions, cardiac arrest, and severe metabolic acidosis and he was about to go for coma (Henry et al., 2020). Though the events of hemlock water-dropwort poisoning in humans are rare, a total of 13 infant cases have been reported due to this plant between 1900 and 1978 . Among these cases, 70% were reported as fatal cases (https://en.wikipedia.org/wiki/Oenanthe_crocata)

Case report 3: Poison hemlock poisoning

A 59 year-old male farmer was admitted to the Emergency Department for respiratory distress and loss of consciousness after about 3 hours of his unintentional ingestion of two pinches of an unknown plant which was subsequently identified as poison hemlock. His other symptoms were spontaneous extremity movements with muscle spasms, dilation of pupils. He was treated as an inpatient and he showed improvement after a period of about 6 hours. After 2 days of observation the patient was discharged with total recovery (Erenler et al., 2011)

7.3. Invertebrates

7.3.1 Protozoans

Case report 1: This case report deals with the infection of the protozoan parasite viz. Balamuthia mandrillaris. A 69- year-old Caucasian female who presented with a 3-day history of worsening confusion and difficulty with speech. On observation , she was disoriented and possessed expressive dysphasia (language disorder). Motor examination of this patient showed a right arm pronator drift and subsequent cerebellar examination exhibited slowing of finger-nose testing on the left. Even after hospitalization and treatment , she died a few weeks later (Yohannan and Feldman,2019).

Case report 2 : This case report deals with the infection of the protozoan parasite viz. Blastocystis hominis. Qadri et al. (1989) reported that among the patients with the B. hominis infection,46% had symptoms. The most common symptom was abdominal pain (87.9%) followed by constipation (32.2%), diarrhea (23.4%), alternating diarrhea and constipation (14.5%), vomiting (12.5%), and fatigue (10.5%) in that order.

Case report 3: This case report deals with the infection of the protozoan parasite viz. *Vermamoeba vermiformis* (= *Hartmannella vermiformis*). The medical significance and its pathogenic potential as a parasite for humans has been reported from several countries like Spain, Costa Rica, Peru, France, Scotland, Japan and Iran. In 2014, this species was detected in human nasal swabs . Most of the case reports relating to this species are however associated with corneal damage. A keratitis case was reported in a female contact lens wearer, who initially had eye pain, redness, and blurred vision, Interestingly, this species has been reported predominantly in mixed amoebic keratitis cases (Scheid, 2019).

Case report 4: This case report deals with the infection of the protozoan parasite viz. Sappinia diploidea . In 1998, only a healthy 38-year-old man from Texas was found infected by *Sappinia diploidea* However, this infection was subsequently reidentified as it was caused by S. *pedata* . Although his hospitalization, he survived without any long-term consequences. Further, it is not known whether Sappinia diploidea can cause infections in human or animal hosts (da Rocha-Azevedo et al., 2009; https://www.cdc.gov/parasites/sappinia/general.html)

7.3.2 Sponges

Drulia ctenosclera

Case Report 1: Rigid microscopic particles (spicules) from the skeletons of these sponges have been reported to penetrate the membranes covering the eyes and cause serious problems in swimming children as observed in the Municipality of Araguatins (located in the Brazilian state of Tocantins). After a dip in water , the children complained of redness, itchiness, and pain and burning in their eyes. In many affected children, superficial scratches or lesions turned into small nodules or opaque spots on the outermost membranes, as a result of fungal or bacterial infections. Further, it was reported that two children lost sight in one of their eyes (https://revistapesquisa.fapesp.br/en/2013/04/02/the-dangers-of-swimming-in-the-araguaia/)

Case Report 2: Cruz et al. (2013) reported that due to the above sponge species, 83 eyes were affected. Among the affected eyes, the most common lesions were the corneal opacities as observed in 34 eyes and conjunctival nodules diagnosed in 12 eyes. Further, they reported severe visual acuity loss which was detected in seven children with unilateral anterior chamber lesions. Furthermore, spicules of these sponge species were found inside three blind eyes which have been enucleated for diagnostic purposes. Organic enrichment of the water was reported as the key factor, which initiated a cycle of ecological imbalance that provoked the aforesaid human eye disease

Drulia uruguayensis and *D. ctenosclera*

Three patients of 8-13 years old were brought to the eye hospital with ocular injuries during July 2005. These affected people were in the habit of swimming in the Araguaia river, Brazil. They were first examined by ophthalmologists and subsequently underwent diagnostic and therapeutic surgical procedures . On thorough examination, It was found that the conjunctiva or cornea surfaces of the eyes of these patients were affected by sharp materials resembling tiny glass tiny bodies. Further detailed examination of these tiny bodies revealed that these tiny bodies are nothing but the silicious spicules (gemmoscleres) of the freshwater sponges of the species *Drulia uruguayensis* and *D. ctenosclera* present in the said river (Volkmer-Ribeiro et al., 2006). The case reports these 3 patients are given below

Case Report 1 : This relates to a 12-year-old boy who was admitted to the eye hospital for his progressive low vision which he had for about 1 year. Biomicroscopy results conveyed that his left eye had inferior adherent leukoma and he was treated with oral prednisone (40 mg/day) and 1% prednisolone acetate, in addition to oral prednisone (40 mg/day).Subsequently Sphincterectomy (with Nd:YAG laser) was performed to remove the anterior synechia and clear visual axis. After this treatment, he showed an improvement in the visual acuity. As he developed cataract in his left eye, he then underwent cataract . On March 2, 2006, the fundus examination revealed a normal posterior pole (Volkmer-Ribeiro et al., 2006).

Case Report 2 : It related to a 13-year-old girl who reported slow and progressive decrease in her visual acuity which had begun some four months ago. He was given a classical treatment to toxoplasmosis . Subsequently she underwent scleral buckling, pars plana lensectomy and vitrectomy in her right eye (Volkmer-Ribeiro et al., 2006).

Case Report 3: This relates to a 8-year-old boy who complained low vision in his right eye for six months. After thorough examination, he was treated with 1% tropicamide , 0.3% ciprofloxacin , and 1% prednisolone acetate in his right eye. However , as he complained worsening of vision and inflammatory reaction, he was given oral prednisone subsequently. Finally, he also underwent cataract extraction in his right eye. After surgery, he further developed an inflammatory pupillary membrane. So a new surgery was to be programmed by the ophthalmologists for this patient (Volkmer-Ribeiro et al., 2006).

7.3.3 Nematode (Guinea worm)

Case report 1 : An 8-yer old girl of rural area came to the hospital with the main complaint of pain in her left leg for the past 15 days and an ulcer on the left shin (front of the leg below the knee) . These symptoms according to her were recurring for the last 4 years. Even after the said ulcer was treated with antibiotics initially and then with multidrug therapy (MDT) for leprosy, the ulcer was found to recur. On further examination of the ulcer with X-ray, it revealed a linear calcification ending in the ulcer. Radiological features confirmed the presence of a calcified Guinea worm in the place of linear calcification. As the lesion was chronic and recurrent, the worm was removed surgically and the skin was sutured in layers . Subsequently, the wound healed completely and thereafter there was no recurrence even after 3 years of observation(Gulanikar, 2012)..

Case report 2 : A 25 year old woman complained of nodular lesion on her left side abdomen for 5 to 7 years this was associated with pain and discharge for 3-4 days. As the lesion was also progressive, she was subject to dermatological examination which confirmed the presence of a large nodular lesion of size 5 cm in diameter. The lesion was subsequently excised and the whole specimen was sent to histopathological analysis. The report confirmed the presence of a calcified Guinea worm (Gulanikar, 2012)

7.3.4 Annelids

Hirudo medicinalis

Case Report 1: A 42-year-old woman was admitted to an emergency room with painful and itchy lesions on her feet and legs. She developed her knee and leg pain due to 14 leeches which were adhered onto both her legs. There were also left rubor, tumor and some other areas of itching on leech-adhered regions. Physical examination of this patient showed that the right popliteal region and the same leg posterior area lesions were found to be erythematous, edematous, and plaque-like shaped, as well as ecchymotic and hemorrhagic areas at the centre of the lesions (Lok et al., 2013).

Case Report 2: A 14-year-old affected female girl with abnormal vaginal bleeding or irregular spotting was admitted to Ghaem Educational, Research, and Treatment Center in Mashhad, Northeast Iran. She had menorrhagia with clot formation which had begun one month before the referral (Hasanzadeh,2019)

Limnodrilus hoffmeisteri

Case report: Liu et al. (2017) reported on an incidental human infection with this species in the nasal cavity of a 25-year-old man residing in Zhangjiakou city, Hebei province, China . He presented nose bleed, severe itching, continuous sneezing, and rhinorrhea. A lot of worms of this species were found in the nasal discharge of this patient.

7.3.5 Arthropods

Lethocerus delpontei (= *Belostoma delpontei*)

Case report: Seven cases of patients bitten by giant water bugs have been reported. The victims experienced intense, excruciating pain and case 1 manifested hypoesthesia in the forearm (Haddad Jr., et al., 2010).

7.3.6 Molluscs

The disease schistosomiasis which causes intense abdominal pain and discharge of blood in the stool or urine of the affected people is due to the freshwater snail carrying parasitic worms. The WHO estimated deaths between 20,000 and 200,000 due to the said disease (https://www.sciencealert.com/here-s-a-list-of-the-top-15-deadliest-animals-on-earth)

7.4. Fishes

7.4.1 Cartilaginous Fishes

Freshwater stingrays

The freshwater ecosystems of South America are known for the envenomations caused by freshwater stingrays. The fear of injuries by freshwater stingrays in the Paraná River Basin (includes the greater part of southeastern Brazil, Paraguay, southeastern Bolivia, and northern Argentina.) has been existing for centuries. In the Brazilian stretch of the Paraná River, stingray envenomations have been reported to be present for the past 20 years (Haddad Junior et al. , 2013a).

Case report 1: Haddad Junior et al. (2013a) reported that the freshwater stingrays caused injuries for the fishermen of Araguaia-Tocantins River Basin (Brazil) mainly during the months of July and August when thousands of people visit this basin for recreational activities. Victims are believed to be stung on their lower limbs during the day, while they are swimming or moving through the water from one sandbar to another. Though, the accidents have been reported decreasing, sporadic cases occurred in amateur and professional fishermen. Accidents usually occurred in the water, when the cast nets were used. Fish unloading or maintenance of boats also resulted in accidents

Case report 2: In the Paraguay River Basin (which is straddling the borders of Brazil, Argentina, Bolivia, and Paraguay) , the freshwater stingray envenomations are very common among the river dwellers, and professional fishermen who are believed to be wounded in their lower limbs during fishing or recreational activities along the rivers. Children are also affected when they step on stingrays in the rivers. Improper handling of hook-captured stingrays cause injuries in amateur fishermen. Tourists and animals such as horses and cattleare also said to be affected by the stingrays while they are bathing in river beaches or crossing rivers. Studies made in the region of Corumbá town, Mato Grosso do Sul state,(Brazil) revealed that the period in which the stingray accidents were more frequent was found to be during the winter i.e. between July and September. Of the 18 envenomation cases observed

, 90% occurred during the day time and males, aged between 29 and 46 years were the most affected lot. The species Potamotrygon motoro and P. falkneri were largely responsible for the human injuries (Haddad Junior et al. , 2013a).

Case report 3: This report deals with the envenomations caused by freshwater stingrays among 132 people (103 mean and 29 women) of Juruá Valley, Acre, Brazil. Accidents occurred mainly during leisure time bathing (52%) and fishing (38%) activities. About 90.0% of the envenomaitons occurred during the daytime , especially in the afternoon (53%) and the largest number of cases (42%) occurred in their working places viz. rubber plantations. Symptoms of the injured population were reported to be severe pain, edema, erythema, necrosis, and ulceration of the wound. The most affected parts of the stingray envenomations have been reported to be lower limb (97%), upper limb (2%), and the trunk (1%), more specifically the hip. The treatment for the injuries in these people are largely based on folk remedies. As accidents are often neglected by these people due to low lethality., educational activities and prophylactic measures need to be carried out with first aid and late measures (da Silva et al., 2020).

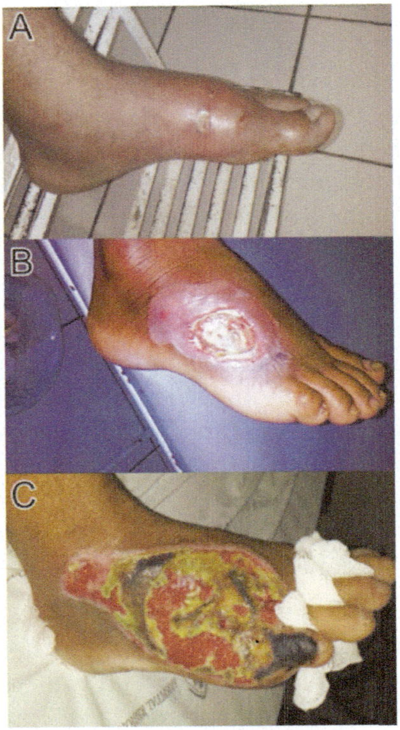

Injuries caused by freshwater stingrays.
Image credit: Dr. Aline Nayara Poscai (Reproduced with permission)

Case report 4: This report deals with 2 stingray envenomation cases associated with aquarium hobby. During the period of this report, small freshwater stingrays were imported to Europe and sold as aquarium fish because of their easy maintenance. Among the two victims, one was an experienced 45-year-old French hobbyist and the second person was a 35-year-old German pet store assistant. Both the victims complained of severe local pain with minimal systemic manifestations. However, after the after symptomatic treatment, both patients recovered fully (Brisset et al., 2006).

Case report 5: The present case involved a large freshwater stingray belonging to the species *Potamotrygon motoro* occurring in the Araguaia River in Tocantins, Brazil. In 2014, a 41-year-old Caucasian male was stung by this species on the medial face of his left foot. This accident resulted in a deep puncture wound with copious bleeding and excruciating pain that quickly spread to the distal part of his thigh (da Silva Jr. et al., 2015).

Case report 6: This report deals with the riverside inhabitants of the Amazonian and Midwest regions of South America who were the main victims of freshwater stingrays. The highest increase in the number of stingray accidents occurred among tourists from other regions. Though the stingray envenomations are generally associated with inflammatory manifestations, such as swelling and erythema, these manifestations often progressed to necrosis which is considered to be potentially tetanogenic. During the study period, Torrez et al.(2015) reported 2 cases of tetanus after freshwater stingray injuries. The first case related to 51-year-old men who had tetanus and recovered without sequel and the second case was with a 67-year-old men who had severe tetanus and a deep, necrotizing soft-tissue infection which resulted in death (Torrez et al., 2015).

Case report 7 : This report deals with stingray envenomations met by 19 people of the municipalities of Santarém and Aveiro (Pará State), Brazil. Of the 19 people 13 were injured while in fishing and 6 in other riverine activities. Fifteen people reported that their injuries due to beach based stingrays and 14 people reported that these injuries occurred during Sept-Nov. While 18 reported severe pain 12 reported formation of ulcers in the later stages of stingray envenomation (Abati et al., 2017)

7.4.2 Teleost Fishes

Catfish spine envenomation

Case report 1: During 2009-2011, two cases of freshwater catfish envenomation were referred to the Noor and Ali Asghar Hospital, Isfahan University of Medical Sciences, Iran. These stings occurred by immersion of hand into the catfish aquarium. While the first patient who was a 42-year-old man complained of a wound on the back of right hand , the second man of 35-year-old had a wound on the palmar area of base of left third finger. On examination these patients showed laceration, bleeding, oedema, and erythema in the area of bites. However, thy had not developed systemic symptoms and signs. The affected limbs were immersed in hot water of temperature below 50 °C which led to a gradual reduction and elimination of pain in about 30 ms. Secondary infection was prevented with the administration of systemic antibiotics infusion of cefazolin 1 g intravenous every 6 h and tetanus toxoid 250 IU intramuscular . Wounds were left open to heal by secondary intention. Both the patients were subsequently observed for incidence of systemic symptoms for 24 h and then they were discharged from hospital with the prescription of oral antibiotics. (Dorooshi, 2012).

Case report 2: A 40-year-old recreational fisherman was admitted to the emergency department 3 h after sustaining an injury to his left hand while he was trying to unhook a freshwater channel catfish , *Ictalurus punctatus*. The spine was found imbedded in the volar aspect of the hand. As it was not possible to remove the spine manually, he cut the spine and removed it. He complained of of severe left hand pain and numbness to the elbow. Initial examination revealed the presence of a 4 X 3-mm foreign body imbedded in the soft tissues midway between the fourth and fifth metacarpal bones. A radiograph of the hand displayed a 2-cm-Iong foreign body with dense circumferential cortex . Initial treatment consisted of warm water immersion for 30 min and tetanus prophylaxis followed by pain management with meperidine/promethazine, and I g intravenous cefazolin. Subsequently , the spine was removed with eventual core excision. The patient was discharged with the prescription of a 7-day course of cephalexin 500 mg four times a day and 48-hr outpatient clinic follow-up. After a four-week clinic follow-up, that the wound was healing well but slowly. At a 6-month telephone follow-up, the patient reported that the wound had healed completely (Blomkalns and Otten, 1999).

Case report 3: Envenomation and associated injuries due to the spotted catfish *Pseudoplatystoma corruscans* and striped catfish, *P. reticulatum* have been reported by 38 fishermen of Miranda town and 111 fishermen of Corumbá town of Brazil. All these fishermen reported lacerated lesions , edema, erythema, radiating pain to the root of the limb, paresthesias and necrosis. A few patients also reported fever and cardiac arrhythmias. These manifestations were believed to be associated with envenomations caused by the toxins in the stingers of the fish or secondary infections (de Aquino et al., 2016).

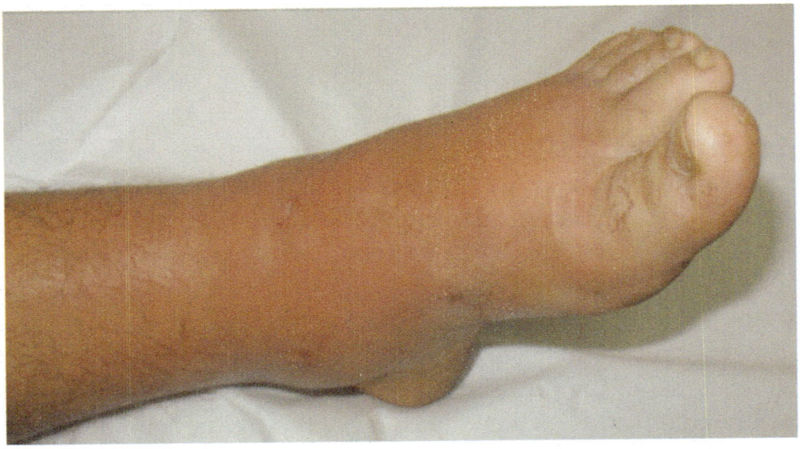

Wound caused by the *Pseudoplatystoma corruscans* catfish
Image courtesy: Vidal Haddad Junior (CC)

Case report 4: Serious infection following catfish spine-related injuries was reported by two patients. On examination, the skin of these patients showed the presence of pathogenic bacterial species of *Vibrio, Aeromonas hydrophila,* Enterobacteriaceae and *Pseudomonas* species. These patients were recommended Irrigation, exploration, and culture of the wounds as well as immunization against tetanus. Wound care and antibiotics were recommended for the patients with normal host defense mechanisms. For moderate to severe infections, a tetracycline and a broad-spectrum, beta-lactamase-stable beta-lactam antibiotic, or a tetracycline, a beta-lactamase-stable penicillin, and an aminoglycoside were prescribed (Murphey, et al., 1992).

Case report 5: This report deals with an young angler of New Port Richey (Florida) who was stabbed by a catfish in his chest. He was airlifted to the hospital and rushed into emergency surgery to remove a hooked catfish barb which is a serrated spine of 5 cm length from his chest. As the needle-like spine was found broken off inside the boy, the boy had breathing trouble and his chest began to swell. This led the boy to St. Joseph's Hospital in Tampa (Florida) for urgent care and treatment due to his suffering shortness of breath (https://www.tampabay.com/news/latest/2022/06/22/boy-stabbed-by-catfish-in-new-port-richey-expected-to-fully-recover/)

Case report 6: This report deals with the injuries caused by catfish to the fishermen community of Pantanal, a flooded area in Midwestern Brazil. Of the total people (47) affected, 14 fishermen had lesions and 9 fishermen showed ulcers. The species of catfish responsible for trauma and envenomations were found to be mainly spotted catfish (*Pseudoplatystoma corruscans*) and striped catfish (*P. reticulatum*) ; and other catfish such as flatwhiskered catfish, *Pinirampus pirinampu*, duckbill catfish, *Sorubim lima* and porthole shovelnose catfish, *Hemisorubim platyrhynchos*. Awareness programmes such as leaflet distribution and discussions significantly decreased the injuries say only two in the next year. Further, such campaigns educated fishermen about the prevention of catfish envenomation and first aid for injuries (Haddad Jr., 2018).

Case report 7: A 52-year-old Chinese-American man was admitted to the emergency wing as complained with erythema and swelling in his right arm. He had these symptoms for about 10 days after skin penetration by a catfish barb. Further, an abscess of his right thumb had undergone incision and drainage which was sent for wound culture immediately. Laboratory reports revealed high white blood count, sedimentation rate, and C-reactive protein. So the patient was treated with intravenous ampicillin-sulbactam and vancomycin and thereby the symptoms improved. Wound culture report showed the presence of pathogenic bacteria viz. *Proteus vulgaris* and *Morganella morganii.* He was subsequently discharged with the prescription of a 10-day course of oral ciprofloxacin and amoxicillin-clavulanate. After a 12-month follow-up, the patient reported that the wound had healed completely (Huang et al., 2013).

Case report 8: This report deals with the catfish injuries caused by driftwood catfish *Centromochlus existimatus* and *C.* heckelii during bucket bathing. Of the 27 people involved, 17 people reported injuries and 3 of them had been injured 2 to 3 times. Interestingly all the catfish envenomations had occurred at dusk or early night. The occurrence of stings by the aforesaid driftwood catfish species to people bucket bathing in the Brazilian Amazon was found to be generally high which is largely due to high chances of catching these small catfish during random throws of a bathing bucket into the Negro and Solimo^es rivers in Brazil (Sazima et al., 2005).

Case report 9: This report deals with an attack of 2 m catfish on a woman who was swimming in a lake in Straubing-Bogen in southeastern Bavaria. She had a bite in her leg due to a huge flesh-eating 2 m long wels catfish. As per an expert of this locality such attacks are common especially when the male fish is protecting a pair's eggs (https://www.thelocal.de/20160825/6-foot-catfish-attacks-woman-in-bavarian-lake/)

Case report 10: This report deals with 78% of the 100 accidents among the fishermen of Coxim and Corumbá municipalities, state of Mato Grosso do Sul, Brazil during December 2008 and October 2009. The catfishes of the genus *Pseudoplatystoma* were responsible for 24 cases; and genera *Pimelodus, Pimelodella* and *Rhamdia* for 23 cases. The symptoms possessed by the patients were wounds (2 cases); edema (8); erythema (3); tachycardia (1); and numbness (1). Though the envenomations used to be more frequent on the hands, the present data showed significant number of accidents (35% of cases) was with lower limbs. In all the patients, the sting of the catfish was found to penetrate their tissues and the barbs released toxins which flew into the wounds causing violent pain for more than six hours, with the possibility of further bacterial infections with deep tissue necrosis due to *Staphylococcus* , *Streptococcus, Vibrio vulnificus* and *Aeromonas hidrophyla*. For all these complications, the presence of fragments of the stinger in the wounds was the causative factor (Silva et al., 2010).

Case report 11: A 51-year-old Malay man was brought to the emergency department because of his left hand pain caused by a puncture wound from the spine of a catfish species, *Pangasianodon hypophthalmus* . Further information is not available in this case (Ahmad et al., 2013)

Case report 12: This report deals with the death of a Brazilian fisherman during 2008. His death was mainly due to a fatal heart perforation caused by a catfish spine when he had held a net full of catfish (species unknown) close to his chest. The onlookers saw the affected fisherman falling into the water and tried in all possible ways to rescue him. But he passed away after a few minutes (https://pondinformer.com/are-catfish-dangerous/)

Case report 13: The catfish species *Mystus pelusius* and *Parasilurus triostegus* occurring in the freshwater systems of southern Iraq have been reported to be responsible for several attacks on humans. The persons affected displayed harsh pain, engaged foreign bodies, infection, respiratory constrained, arterial hypotension, and cardiac dysrhythmias. The treatment for such incidences included hot water immersion, analgesia, injury examination, and prophylactic antibiotics (Jawad, 2021)

Case report 14: This report deals with 5 cases of a total of 127 injuries due to catfish attack in the Brazilian coast during a ten-year period . Radiological examination (a simple X-ray) of 5 affected patients revealed the retention of fragments of the stinger in the wounds. Among these five patients, three patients had been stung in the feet, and two had been stung in the hands. However, all the 127 affected cases showed intense local pain and inflammation. They were treated with hot water immersion which improved some symptoms. However, the 5 cases with the retention of fragments of the stinger in the wounds, local inflammation and pain persisted for more than 24 h. These patients returned to the hospital 1-5 days after the initial envenomation. (Negreiros et al., 2016).

7.5. Amphibians

7.5.1 Toads

1) Case report relating to toad venom poisoning

A 40 year old man was admitted to the emergency department with complaints of abdominal pain, nausea, vomiting and diaphoresis. He admitted that he had ingested three sex (aphrodisiac) pills which were suspected to be toad venom. His other symptoms were mild epigastric tenderness and lethargic condition. ECG showed sinus bradycardia with first degree atrioventricular block and a chest x ray displayed an early right basal infiltrate. Subsequently serum and urine toxicological screening and haematological investigation were also made and his condition however, deteriorated rapidly, with severe respiratory distress. Even after tracheal intubation, (insertion of a tube into the patient's body part, particularly the trachea for ventilation), his condition deteriorated alarmingly, with severe bradycardia. Subsequently he developed ventricular tachycardia. Despite all life supporting measures ,the patient died (Gowda et al., 2003)

2) Case report relating to ingestion of toad eggs

A 19-year-old woman was admitted to the emergency department with complaints of numbness in mouth, nausea and vomiting. She admitted that she had ingested cooked toad eggs about an hour back. On examination, she had mild epigastric tenderness. Though haematological investigations did not show any adverse conditions of the patient, ECG showed a complete atrioventricular block. Further gastrointestinal absorption of toxin was prevented with gastric lavage and activated charcoal. Hyperkalaemia was also managed with insulin, glucose water and sodium bicarbonate. She was subsequently admitted to the intensive care unit and she was administered Fab fragment with an initial dose of 80 mg and another dose of 40 mg after 1 h. Due to these clinical procedures, her potassium and digoxin level declined gradually and her condition improved noticeably. She was discharged in fair condition on the fifth day (Kuo et al., 2007)

7.5.2 Frogs

Phyllomedusa bicolor

Kambo (cane toad secretion) and its adverse effects : Kambo is a traditional ritual in which the poison of the giant monkey frog, *Phyllomedusa bicolor* Is used to treat various health conditions and purify the body (https://www.medicalnewstoday.com/articles/323689). Hesselink (2018) reported that the misuse of kambo or the use of wrong frog may lead to severe adverse events or even death. Very often confusion prevails in the differentiation of the kambo yielding giant monkey frog, *Phyllomedusa bicolor* and the toxic toad of the genus *Chaunus*. This case report relates to two deaths (one in Pindamonhanga (São Paulo) during 2008 and the other in Chile during 2009) supposedly caused by kambo. As further information is wanting , no conclusion was drawn relating to these deaths. However , it was suggested that some fatal cases in the USA were due to drinking of too much water before and/or after the kambo ritual. The adverse conditions due to 'purification ritual' include the formation of scars on the shoulder of the patient; and heart throbbing, sweating and nausea arises that leads immediately to severe vomiting within minutes of the use. However, these effects disappeared after a few more minutes. It is also suggested that the bioactive peptides present in kambo may also lead to tachycardia, dizziness, nausea, vomiting and sedation. Furthermore, the presence of phyllocaerulein, phyllomedusin and phyllokinin of kambo has also been reported to cause vasodilatation, hypotension, and reflex tachycardia (Hesselink,2018).

Case report 1: A 44-year-old woman drank six liters of water after a kambo ritual. Within 5 minutes she was reported to be dizzy and confused, and after half an hour she seemed well again. Subsequently she drunk a recommended amount of four liters of water. Three hours later she started vomiting and shivering and developed generalized muscular cramps. Thereafter she drunk another 2 liters of water. Nine hours after the start of the ritual, she was found to be nauseated, confused, agitated, dysarthric, tremulous and ataxic and painful muscular cramps in arms and legs. On the way to hospital a generalized convulsion was witnessed by the emergency physician. Laboratory findings plasma hypo-osmolality (251 mosmol/kg), hyponatremia (116 mmol/L), elevated urine osmolality (523 mosmol/kg) and high urine sodium concentration (87 mmol/L), all pointing towards a SIADH. Treatment with 0.9% sodium chloride at a rate of 100 ml/h and a restriction of water intake was initiated, one day later she was awake and orientated. The case was complicated by rhabdomyolysis. (Hesselink,2018).

Case report 2: A similar case is reported in literature in a 22-year-old man, who also drunk 6 liters of water within some hours. He also he showed progressive psychiatric and neurological signs including restlessness and convulsions (Hesselink,2018)..

Case report 3: A 34-year-old overweight man with a long history of alcohol and Cannabis use was admitted to hospital due to icterus, skin itching, weakness, and pain in the upper abdomen. Patient wanted to stop alcohol and Cannabis use and started Kambo applications once a week to assist the process. Laboratory data: Signs of a transient hepatitis. Patient returned home after 10 days in good condition with normalized laboratory functions (Hesselink,2018). .

Case report 4: A sudden death of a 42-year-old overweight man with signs of coronary pathology was reported dead shortly thereafter. It was suggested that in this case perhaps the hypotensive effects of Kambo could have resulted in reduced myocardial perfusion due totachycardia, perhaps leading to a cardiac arrhythmia followed by sudden death.

There were two reports of prolonged symptoms after Kambo administration, such as nausea, vomiting, flushing and facial swelling. These are clearly and directly the pharmacological effects of Kambo. Due to a number of factors such as the purity of Kambo, the concentration of the bioactive peptides, the short 'shelf-life time' of the active components in climate zones with high humidity and temperature, as well as the amount and size of the dots, the bioavailability of the peptides may fluctuate considerably, leading to a somewhat unpredictable intensity and time course of the symptoms (Hesselink,2018).

Case report 5: Kambo is a ritual based on the 'application of a toxin' and it was reported as the psychotic decompensation of a Belgian woman, who claimed to be a certified shaman, practicing healing through the utilization of Kambo from once per month to up to nine times per month. In the past the police were repeatedly called by the patient about non-existing rapes and shootings in her community. She clearly had a history characteristic for paranoia, including bizarre delusions, but the paper was unclear about any causal relation between the start of Kambo treatment and the evolution of the psychiatric symptoms. She was brought to the hospital in an incoherent mood, suffering from various psychiatric symptoms: Paranoia, anxiety, bizarre delusions, labile mood, and panic attacks. She normalized after the start of risperidone treatment. However, this case report was not a convincing one and it did not support a causal relation between Kambo and the paranoid syndrome. It might very well be that prior to the taking of Kambo the so called 'shaman' already suffered from psychiatric disturbances (Hesselink,2018).

Case report 6: A fatal case was reported in Poland and it related to the death of a 30-year-old women, who took Kambo. She first started to vomit and subsequently lost consciousness. She was transported by ambulance to the local hospital, where she was connected to an artificial respirator and subsequently died. Apparently, there were also neurological symptoms of intracranial high pressure, cerebral edema and electrolyte disturbances. Therapy, probably with high dose of corticosteroids and supplementation of the electrolyte deficiency did not result in improvement After a few hours, the woman was transferred to the intensive care unit, where she unfortunately died. More details were not provided, and especially lacking was the information whether the patient in question perhaps had an excessive water-intake, resulting in severe hyponatremia before or after the ritual, a recognized cause for SIADH and its cerebral complications (Hesselink, 2018)..

Case report 7: Three additional cases of hyponatremia were reported in the UK, USA and in Australia. The last one was related to a client dieting on distilled water. All people were saved by sodium infusions and appropriate care in the hospital (Hesselink, 2018).

7.5.3 Newts

Case report : A 29-year-old man swallowed a 20-cm (about 8 in) newt on a dare. Within minutes, he complained of tingling of his lips, which progressed to numbness and weakness of his entire body over the next 2 hours. Although he stated that he thought he was going to die, he refused to be transported to a hospital. He then collapsed and went into cardiac and respiratory arrest. He was resuscitated and given respiratory support, but his pupils were dilated and fixed. A brain study confirmed that he had no brain function, and he was pronounced dead later the same day (https://www.poison.org/articles/poisoning-from-amphibians-204#:~:text=Some%20of%20the%20potent%20toxins,to%20spread%20 Salmonella%20to%20humans)

7.6. Reptiles

7.6.1 Freshwater snakes (water moccasins/cottonmouths, and rattlesnakes)

Annually about 5000 snake envenomations are reported to the American Association of Poison Control Centers annually. Among the venomous snakes, the pit vipers viz. copperheads, rattlesnakes, and water moccasins (cottonmouths) are responsible for the most snakebites in the United States. While copperhead bites occur more commonly in the southeastern United States rattlesnake bites in the west. Further , copperhead snakes which are also found in southern Canada and northern Mexico are believed to be responsible for almost half of all venomous snakebites in the USA. Southern and western states of USA with warmer climates such as Texas, Oklahoma, Florida, California, Arizona, Louisiana, Georgia, and North Carolina have shown the greatest number of venomous snakebites . Snakebites are more common in the summer months.

Clinical manifestations after the rattlesnake envenomation have been reported to be more severe in patients. However, fatalities are said to be rare (Seifert and Ruha, 2022). Another report says that about 50 % of all snake bites in the United States are on lower extremities and there were 255 cottonmouth envenomations during 2017, Among them, 242 people were treated by healthcare professionals; 122 patients showed moderate symptoms ; and 10 patients suffered with severe symptoms. However nobody died (https://a-z-animals.com/blog/how-many-people-do-cottonmouths-water-moccasins-bite-per-year/)

Water Moccasin Snakes

Case report 1: In the USA, the majority of venomous snake exposures are due to pit vipers viz. There rattlesnakes , copperheads and cottonmouths (water moccasins). The cottonmouth bites are however, reported less frequently than other pit viper bites. During 2013- 2017 , 31 cottonmouth envenomations have been reported. Most of these bites occurred in children of 7–12 years. Swelling was found to be the most reported clinical manifestation which occurred in all patients. Gastrointestinal symptoms were observed in 19% of patients, and coagulopathy in 19% . While 19 patients needed hospital stays of more than a day 24 hours, antivenom treatment was given for others . (Domanski et al., 2020)..

Case report 2: In the United States of America, during 2006-2015 , the number of people exposed to the freshwater pit vipers (Family: Viperidae; subfamily Crotalinae) s viz. copperhead, rattlesnake, and cottonmouth/water moccasin were 15,107, 11,530 and 2300 respectively. The fatalities recorded by these snakes were 2,19 and 1 respectively (https://www.ebmedicine.net/topics/toxicology-environmental/snake-bite-antivenom). Morton and Peter (2015) reported that 99% of snake envenomation cases in the United States of America were due to the freshwater pit vipers. Further, in 2012 alone , 1,766 cases of copperhead snake envenomation were reported and 382 of these cases occurred in people of 19 years or younger. The high risk of mortality and morbidity in children is due to their smaller body size in relation to the dose of venom. Richardson et al.(2005) reported that in the United States of America approximately 8000 crotaline (pit vipers) bites were recorded annually and deaths are only 10-15.

Case report 3: This report deals with a pediatric case of suspected venomous bite. The patient was an 8-year-old boy who felt a bite when he was playing in the grass in the yard of his home at about 7:15 pm. Immediately after the bite, bleeding, swelling, and bruising of the boy's thumb began. After giving the boy a dose of diphenhydramine, he was taken to the emergency department (ED) . In the mean time the snake was identified as the copperhead , Agkistrodon contortrix . In the hospital ,one vial of Fab antivenin (CroFab) was first administered . As swelling and joint pain over the boy's right thumb worsened , he was transferred to a metropolitan pediatric ED where 3 vials of the recommended dose of CroFab were administered. As there was considerable improvement he was discharged after about 8 hrs (Morton and Peter, 2015)

Case report 4: This report also deals with a pediatric case of suspected venomous bite. When a 9-year-old boy was fishing with his father, he observed a sharp, stinging sensation on his left hand. This was suspected to be a snake bite particularly of copperhead. Though bleeding was minimal initially, bruising and swelling starting developing up to his elbow. Due to the progress of swelling and the appearance of a puncture wound, he was admitted in a metropolitan pediatric ED. He was first administered 2 mg/kg of methylprednisolone sodium succinate for injection. As the swelling continued to decrease, and his lab results showed normal range, he was discharged (Morton and Peter, 2r015).

7.6.2 Rattlesnakes

Richardson et al.(2005) reported that mortality from rattlesnake envenomation in the United States is rare. The fatalities due to these snakes are largely due to their direct intravascular envenomation and anaphylaxis and bites to the head and neck of the individual

Case report 1: This case deals with the bite on the face of an individual. As the patient showed marked coagulopathy and prolonged airway compromise, he required 9 days of nasotracheal intubation. As the bite site did not develop necrosis, only local wound care was required (Lewis and Portera Jr, 1994).

Case report 2: This report deal with a pediatric case of severe facial envenomation which required emergent intubation and antivenom administration. A 14-month-old female child had a bite in her right upper lip by a Southern Pacific rattlesnake (*Crotalus viridis helleri*) . As the child developed rapid swelling and ecchymosis ,she was airlifted to a pediatric tertiary care hospital. Within 3 hours of her admission, her stridorous respirations were complicated by significant facial and oropharyngeal edema which required emergent orotracheal intubation. She was administered with 16 vials of FabAV [Crotalidae Polyvalent Immune Fab (ovine) antivenom] over the next 24 hours. The child started gradually improving and she was successfully extubated 5 days later. A 3-month follow-up revealed no significant facial abnormalities in her (Lewis and Portera Jr, 1994).

Case report 3: On 5 September 2018, a 34-year-old male from the city of Bonfim, Brazil was envenomated by a snake on his left ankle . The snake was subsequently identified as a rattlesnake. The patient was however admitted in the hospital 3 h after the snakebite. He had symptoms such as blurred vision , moderate local pain, and mouth dryness. Further, he showed signs of sluggishness and neurotoxicity. Local examination displayed two deep bleeding fang marks, redness, bruising, and swelling at the site of the bite. However, the patient had a fair general condition . He was administered intravenous (iv) injection of 200 mg of hydrocortisone and 10 mg of piriton intravenously (iv) to control allergic reactions, and 70 mg of anti-inflammatory diclofenac. Due to the lack of the specific serum in this hospital he was shifted to the intensive care unit of the nearby big hospital where he received adequate heart rate and blood pressure monitoring, pulse oximetry, and oxygen. To avoid further allergic reactions and pain, he was hydrated with with 1.5 L of physiological solution (0.9%) supplemented with hydrocortisone (200 mg) and metamizole (1 g). Finally he was administered 16 ampoules of crotalic antivenom. Antibiotic therapy with cephalothin (1 g each 6 h) was also prescribed. On the first day, the patient developed hemodynamic instability and he was provided support and vasoactive drugs in order to maintain his blood pressure. In order to control his acute kidney injury , ventilatory parameters were adjusted and sodium bicarbonate and calcium gluconate were administered (iv) for controlling acidosis. Even after these treatments, the patient persisted with hemodynamic instability which lead to 3 cardiac arrests. Finally he passed away on the night of the same day of admission (Medeiros et al., 2020).

Case report 4: A 32-year-old man who is taken to the emergency department reported a snakebite possibly by a prairie rattlesnake. He had severe pain at the site and progressive swelling up his arm. Local examination of the bite site showed tissue damage and swelling. He was administered with antivenom and he developed an anaphylactic response which responded to initial drug cessation, antihistamines, steroids, and epinephrine use (https://www.acep.org/tox/case-files/27_Rattlesnake%20Envenomation%202.pdf)

Case report 5: A 58-year-old woman from Santa Monica Mountains (Southern California) was admitted in ED after sustaining a rattlesnake bite on her left lower leg. As she had hypotension, tachycardia, tachypnea, and facial swelling, administration of oral diphenhydramine did not yield desired results. She was subsequently given intravenous (IV) fluids and she was transported emergently to the city- based Cedars-Sinai Medical Center for further treatment (Lefkowitz et al., 2012)

Case report 6: A 51-year-old man was admitted in a local community hospital within 2 hours of a possible snake bite on his right forearm. The snake was subsequently identified as Eastern Diamondback rattlesnake (*Crotalus adamanteus*). He was administered with 10 vials of antivenom and 4 units of Fresh Frozen Plasma (FFP) and he was also intubated before his transfer to a tertiary care academic medical center. He had symptoms such as anxiety and hypertension. Due to his persistent coagulopathies and development of a gastrointestinal bleed with melenic stool, the patient was administered with antivenom, FFP, and cryoprecipitate on the first day. On the 2nd day , his coagulation parameters normalized and clinical bleeding also ceased and this resulted in the discontinuation of antivenom and management of hypofibroginemia with cryoprecipitate. However, later that day, his coagulation parameters worsened and antivenom had to be reinitiated along with FFP and additional cryoprecipitate. His coagulopathy resolved on the 14th day after the patient had received a total of 45 vials of antivenom (CroFab®), 21 units of FFP, 7 units of Packed Red Blood Cells (PRBC), and 27 units of cryoprecipitate. The final 4 vials of antivenom were also administered on the 12th day due to concern for worsening coagulopathy. Further, the patient's hospital course was complicated by methicillin-susceptible *Staphylococcus aureus* pneumonia, hypotension and acute kidney injury from 10th to 16th day. Suitable treatments were given for all these ailments. On the 17th day, after the completion of therapy for pneumonia, a tracheostomy was done. Decannulation occurred on the 28th day. Eventually the patient›s renal recovery was quite sufficient to discontinue hemodialysis, and he was finally discharged home on the 30th day (Roberts et al., 2017)

7.6.3 Crocodiles

Crocodiles have been reported to be responsible for the most human deaths in Africa. According to the FAO of the United Nations, about 1000 human deaths occur due to these animals (https://www.sciencealert.com/here-s-a-list-of-the-top-15-deadliest-animals-on-earth)

a) *Melanosuchus niger*

Case report: This case report relates to an 11-year-old girl who was attacked by this species in the Mamoré and Novo Pacaás Rivers, in the Brazilian Amazon.

As the girl died after this attack, her body was finally traced. This body revealed different cutting/perforating wounds of about 2–3 cm in length and all these wounds were more evident in the thorax, lower limbs, and buttocks. There were no large lacerations or tearing of extremities. The victim also sustained a fracture of the right femur with animal's teeth marks on the thighs. As the lungs of the dead body were filled with water, the cause of death was confirmed as drowning (Haddad Jr. and Fonseca,2011).

b) *Crocodylus acutus*

Case report: In May 2007, two children were attacked and killed by this species (one in Mexico just south of Puerto Vallarta and one in Costa Rica). On August, 2014, a man and a woman who were swimming in a canal in Florida were bitten in the shoulder and the hand by this species. This was the first documented wild crocodile attack on humans in Florida. A total of 36 American crocodile attacks on humans have been reported from 1995 to 2017 in the Cancun area of southeastern Mexico (https://en.wikipedia.org/wiki/American_crocodile).

7.6.4 Alligators

Case Report: This case report deals with a 43-year-old male who was infected by the bacterial disease leptospirosis due to his contact with the carcass of an alligator. This patient had symptoms such as malaise, upper respiratory infection and neck pain. Initially he was prescribed amoxicillin (500 mg three times per day) , clemastine fumarate/phenylpropanolamine hydrochloride, prochlorperazine , acetaminophen and acyclovir . When he further complained hallucinations, disorientation, fevers of 104 to 105 °F, severe headaches, a bright red rash on his chest, bloodshot eyes, and a bright cherry red tongue he was admitted to the hospital for further evaluation. The physical examination revealed his swollen red eyes, red oral pharynx and the glossy red tongue. As the patient 's presentation was consistent with leptospirosis, he was further treated with 500 mg of ampicillin IV intravenously every six hours initially on hospital day one, and then the dose was increased to 2 g of ampicillin IV every six hours on hospital day two. He was also placed on tobramycin 80 mg IV every eight hours, and then 500 mg of erythromycin IV every six hours was added to his medications on hospital day three. The patient however developed intrapulmonary bleeding with respiratory failure. A bronchoscopy with biopsy was done, followed by intubation as a result of intrapulmonary bleeding with respiratory failure. The patient remained febrile during his course at the hospital. In the medical intensive care unit, the patient was treated with 1 g of ampicillin IV every six hours and erythromycin 1 g IV every six hours. As the patient's respiratory status began to improve he was extubated and transferred to the medical floor for further care . Subsequently, the patient's symptoms slowly improved. He continued hemodialysis with gradual improvement of his renal function. The infectious disease team at the university hospital believed that the patient had leptospirosis and repeated titers were drawn and found to be positive. The patient's hematocrit decreased to 20 g/

dL and he was transfused. Hemodialysis was stopped as the patient's renal function increased (Feuer and Domash-Martinez, 2011).

7.7 Mammals

Hippopotamus

Case report: A total of 11 patients who suffered hippopotamus bite injuries have been reported from Burundi (east-central Africa) in a period of 2 years. Among these patients, nine were males and two were females and the mean age of the patients was 31 years. Most of the injuries were associated with a river or Lake Tanganyika. Of these 11 cases, three met injuries while swimming, two in farming, two in fishing and one while washing clothes. No report on the remaining three patients. Further, seven people were injured during day light and two between sunset and sunrise. All the 11 patients were admitted to the emergency department after suitable resuscitation and initial wound care. The average number of days in the hospital for these patients was 19 days and two patients had readmission. One of the latter was re-admitted twice.

Injury types: The affected people complained of severe wound infections, amputations and permanent disability. The injuries due to hippopotamus bite are therefore considered as major trauma, rather than just 'mammalian bites'. Lower limb injuries were found to be most common in eight people; five had upper limb injuries ; four showed penetrating wounds in the abdomen and chest ; and three patients were with open long bone fractures. Depending on the nature of injuries, four patients were triaged (assigning degrees of urgency to wounded or ill patients) as 'red', four were 'orange' and three were 'yellow'.

Treatment : Among the patients, six got blood transfusion . All patients had surgical interventions in the operating room of the hospital , and three needed laparotomy . Four patients had limb amputation due to traumatic ischemia. All the 11 patients were discharged and there was no mortality However, six people suffered permanent disabilities at discharge (Haddara et al., 2020)

REFERENCES

Abati, P.A.M., Torrez, P.P.Q., França,F.O.D., Tozzi, F.L., Guerreiro,F.M.B., dos Santos, S.A.T., de Oliveira, S.M.S., Haddad,Jr.V., (2017). Injuries caused by freshwater stingrays in the Tapajós River Basin: a clinical and sociodemographic study. *Rev. Soc. Bras. Med. Trop.* 50, 374-378.

Adam,R.D., (2001). Biology of *Giardia lamblia. Clin. Microbiol. Rev.* 14, 447–475.

Ahmed, A.T.A., Rahman, M.M. and Mandal, S., (2013). Biodiversity of hillstream fishes in Bangladesh. *Zootaxa* 3700,283-292.

Alharbi, M.H., Condemine, C., Christiansen, R., LaCourse, E.J., Makaula, P., Stanton, M.C., Juziwelo, L., Kayuni, S., Stothard,J.R., (2019). *Biomphalaria pfeifferi* Snails and Intestinal Schistosomiasis, Lake Malawi, Africa, 2017–2018. *Emerg. Infect. Dis.* 25, 3.

Al-Herrawy, A., Heshmat, M., Abu Kabsha, S., Gad, M., Lotfy ,W., (2015). Occurrence of *Acanthamoeba* species in the Damanhour Drinking Water Treatment Plant, Behera Governorate (Egypt). *Rep.Parasitol.* 4, 15-21.

Allison, https://pubs.nmsu.edu/_circulars/CR678/

Anon., 2009. Securing Water for Ecosystems and Human Well-being: The Importance of Environmental Flows. https://www.iucn.org/downloads/securing_water_for_ecosystems_and_human_well_being.pdf

Anon., 2019. Summary Report – One Health Harmful Algal Bloom System (OHHABS), United States, 2019, https://www.cdc.gov/habs/data/2019-ohhabs-data-summary.html

Anon., 2020. Freshwater Ecosystem: Types, Characteristics, and Animals. https://www.earthreminder.com/freshwater-ecosystem-types-characteristics-and-animals/

Ariza, C., Guevara, D.C., Ubeda,J.M., Cutillas.,C., (1989). Description of four species of the genus *Vannella* isolated from freshwater . *Microbiología SEM* 5, 25-33.

Ashbolt,N.J., (2015).Microbial Contamination of Drinking Water and Human Health from Community Water Systems. *Curr. Environ. Health Rep.* 2, 95–106.

Balczun,C., Scheid, P.L., (2017). Free-Living Amoebae as Hosts for and Vectors of Intracellular Microorganisms with Public Health Significance. *Viruses* 9, 65.

Birkhead,W.S., (1972). Toxicity of stings of ariid and ictalurid catfishes. *Copeia* 1972, 790-807.

Blaise , S., Brun, V.L., Sparsa, A., Delrous, J-L., Bonnetblanc, J-M., (2002). Contact dermatitis with *Hirudo medicinalis*. *Ann. Dermatol. Venereol* 129, 1380-1382.

Blomkalns ,A.L., Otten,E.J., (1999). Catfish spine envenomation: a case report and literature review. *Wilderness Environ. Med.* 10,242-246.

Brand, L.E., Pablo,J., Compton, A., Hammerschlag, N., Mash, D.C., (2010). Cyanobacterial Blooms and the Occurrence of the neurotoxin beta-N-methylamino-L-alanine (BMAA) in South Florida Aquatic Food Webs. *Harmful Algae* 9, 620–635.

Brisset , I.B., Schaper, A., Pommier, P., de Haro.L., (2006).Envenomation by Amazonian freshwater stingray *Potamotrygon motoro*: 2 cases reported in Europe. *Toxicon* 47, 32-34.

Butterfield, B., Bond, A., Meyer, D., DiPerna, G., Peredy, T., (2019). A Case Report of Water Hemlock Poisoning. *J. Maine Med. Ctr.* 1, 12.

Cabral, J.P.S., (2010). Water Microbiology. Bacterial Pathogens and Water. *Int. J. Environ. Res. Public Health* 7, 3657–3703.

Caldeira,R.L., Teodoro,T.M., Jannotti-Passos, L.K., Lira-Moreira, P.M., Goveia, C.D.O., Carvalho,O.D.S., (2016). Characterization of South American Snails of the Genus *Biomphalaria* (Basommatophora: Planorbidae) and *Schistosoma mansoni* (Platyhelminthes: Trematoda) in Molluscs by PCR-RFLP. *Biomed Res. Int.* 6,1-5.

Carvalho, T.I., Klaczko,J., Slobodian, V., (2021). Pectoral-fin glands and delivery apparatus in the catfish genus *Brachyrhamdia* Myers, 1927 (Siluriformes: Heptapteridae). *Pap. Avulsos Zool.*61, e20216174.

Chorus,I.C., Bartram,J., (1999). Chapter 3. Cyanobacterial Toxins. Toxic Cyanobacteria in Water: *A guide to their public health consequences, monitoring and management*. WHO

Chowdhury ,F.R., Ahasan ,H.A.M.N., (2015). Puffer Fish Poisoning: Bangladesh Perspective. P. Gopalakrishnakone et al. (eds.), *Clinical Toxinology in Asia Pacific and Africa, Toxinology*, DOI 10.1007/978-94-007-6386-9_26

Cox, P.A., Banack, S.A., Murch,S.J., Rasmussen, U., Tien,G., Bidigare, R.R., Metcalf, J.S., Morrison, L.F., Codd,G.A., Bergman,B., (2005). Diverse taxa of cyanobacteria produce -N-methylamino-l-alanine, a neurotoxic amino acid. *Proc. Natl. Acad. Sci. U S A* 102 , 5074-5078.

Coyle, C.M., Varughese, J., Weiss, L.M., Tanowitz, H.B., (2012).Blastocystis: To Treat or Not to Treat… *Clin.Infect. Dis.* 54, 105–110.

Cruz , A.A.V., Alencar , V.M., Medina , N.H., Volkmer-Ribeiro, C., Gatta´s,V.L., Luna, E., (2013). Dangerous waters: outbreak of eye lesions caused by freshwater sponge spicules. *Eye* 27, 398–402.

da Rocha-Azevedo,B., Tanowitz, H.B., Marciano-Cabral, F., (2009).Diagnosis of Infections Caused by Pathogenic Free-Living *Amoebae. Interdiscip. Perspect. Infect. Dis.* 2009, 251406.

da Silva, G.A., Poscai, A.N., Casas, A.L.D., (2020). Injuries Caused by Freshwater Stingrays in the Western Amazon: Folk Medicine and Beliefs. *Ethnobiol. Lett.* 11, 1–13.

da Silva, J.P.C.B., Loboda, T.S., (2019). *Potamotrygon marquesi*, a new species of neotropical freshwater stingray (Potamotrygonidae) from the Brazilian Amazon Basin. *J Fish Biol.* 95,594–612.

da Silva Jr., N.J., Ferreira, K.R.C., Pinto,R.N.L., Aird,S.D., (2015). A Severe Accident Caused by an Ocellate River Stingray (*Potamotrygon motoro*) in Central Brazil: How Well Do We Really Understand Stingray Venom Chemistry, Envenomation, and Therapeutics? *Toxins* 7, 2272-2288.

Daly, J.W., Gusovsky, F., Myers, C.W., Yotsu-Yamashita, M., Yasumoto, T. , (1994). First occurrence of tetrodotoxin in a dendrobatid frog (*Colostethus inguinalis*), with further reports for the bufonid genus Atelopus. *Toxicon* 32, 279-285.

Das, M., Dasgupta, S.C., Gomes.A., (2001). Isolation, purification and partial chemical characterization of a lethal factor from common Indian toad, (*Bufo melanostictus*) skin extract. *Indian J.exptl.Biol.* 39, 781-785.

Davis,W., Weil, A.T., (1992). Identity of a new world psychoactive toad. *Ancient Mesoamerica* 3: 51-59.

de Aquino, G.N.R., de Souza, C.C., Haddad,V. Jr., Sabino, J., (2016). Injuries caused by the venomous catfish pintado and cachara (*Pseudoplatystoma genus*) in fishermen of the Pantanal region in Brazil. *An. Acad. Bras. Ciênc.* 88, https://doi.org/10.1590/0001-3765201620150336.

de Carvalho,M.R., Ragno,M.P., (2011). An unusual, dwarf new species of Neotropical freshwater stingray, *Plesiotrygon nana* sp. nov., from the upper and mid Amazon basin: the second species of Plesiotrygon (Chondrichthyes: Potamotrygonidae). *Papéis Avulsos de Zoologia* 51, 101-138.

de Oliveira Júnior, N.G., Fernandes, G.R., Cardoso, M.H.,Costa, F.F., Cândido, E.S., Neto, D.G., Mortari, M.R., Schwartz, E.F., Franco, O.L., de Alencar , S.A., (2016). Venom gland transcriptome analyses of two freshwater stingrays (Myliobatiformes: Potamotrygonidae) from Brazil. *Sci.Rep.*6, 21935.

Dehghani, H., Sajjadi,M.M., Rajaian,H., Sajedianfard,J., Parto,P., (2009). Study of patient's injuries by stingrays, lethal activity determination and cardiac effects induced by *Himantura gerrardi* venom. *Toxicon* 54, 881-886.

Domanski, D., Kleinschmidt ,K.C., Greene,S., Ruha,A.M., Bebarta,V.S., Onisko,N., Campleman,S., Brent,J., Wax,P., (2020). Cottonmouth snake bites reported to the ToxIC North American snakebite registry 2013-2017. Toxic North American Snakebite Registry Group. *Clin. Toxicol. (Phila).* 58,178-182.

Dorooshi, G., (2012). Catfish stings: A report of two cases. *J. Res. Med. Sci.* 17, 578–581.

Drulia ctenosclera. https://revistapesquisa.fapesp.br/en/2013/04/02/the-dangers-of-swimming-in-the-araguaia/

Eberhard,M.L., Cleveland, C.A., Zirimwabagabo, H., Yabsley, M.J., Ouakou,P.T., Ruiz-Tiben,E., (2016). Guinea Worm (*Dracunculus medinensis*) Infection in a Wild-Caught Frog, Chad. *Emerg. Infect. Dis*. 22, 1961–1962.

Eissa, A.E. An overview on the top 10 most dangerous fishes in the world. file:///C:/Users/rsant/Downloads/DangerousFishesArticle-ByAlaaEdinEissa.pdf

Erenler, A.K., Baydin,A., Duran, L., Yardan,T., Turkoz.B., (2011). A case of respiratory failure due to poison hemlock poisoning presented to an emergency department. *Hong Kong J. Emerg. Med.* 18, 235-238

Español , https://www.nrdc.org/stories/freshwater-harmful-algal-blooms-101,2019).

Esterhuizen-Lond, M., Pflugmacher,S., (2020). Chapter 37 - Microcystins as environmental and human health hazards. *Handbook of Algal Science, Technology and Medicine*. p. 591-604

Falade,M.O., Otarigho,B., (2015). Shell Morphology of Three Medical Important Tropical Freshwater Pulmonate Snails from Five Sites in South-Western Nigeria. *Int. J. Zool. Res.*11 , 140-150.

Fayer,R., (2004).*Cryptosporidium:* a water-borne zoonotic parasite. .*Vet. Parasitol.* 126, 37-56.

Fernandez, M.A., Thiengo,S.C., (2006). Susceptibility of *Biomphalaria amazonica* and *Biomphalaria occidentalis* from Manso Dam, Mato Grosso, Brazil to infection with three strains of Schistosoma mansoni . *Mem. Inst. Oswaldo Cruz, Rio de Janeiro* 101, 235-237.

Feuer, B., Domash-Martinez ,T., (2011). Report of case: leptospirosis after exposure to alligator carcass. *Osteopath. Fam. Physician* 3, 23-26.

Fontenelle, J.P., da Silva, J.P.C.B., de Carvalho, M.R., (2014). *Potamotrygon limai*, sp. nov., a new species of freshwater stingray from the upper Madeira River system, Amazon basin (Chondrichthyes: Potamotrygonidae). *Zootaxa* 3765, 249-268.

Gadelha, I.C.N., de Lima, J.M., Batista, J.S., Melo, M.M., Soto-Blanco,B., (2014). Toxicity Effects of Toad (*Rhinella jimi* Stevaux, 2002) Venom in Chicken (Gallus gallus domesticus). *Hindawi*, 851473.

Gowda, R.M., Cohen,R.A., Khan,I.A., (2003). Toad venom poisoning: resemblance to digoxin toxicity and therapeutic implications. *Heart.* 89, e14.

Gulanikar,A., Cutan,J., (2012). Dracunculiasis: Two Cases with Rare Presentations., *Aesthet Surg*. 5, 281–283.

Güémez ,A., Garcia, E., (2021). Primary Amoebic Meningoencephalitis by *Naegleria fowleri*: Pathogenesis and Treatments. *Biomolecules* 11, 1320.

Haddad Jr., V., Fonseca, W.C., (2011). A Fatal Attack on a Child by a Black Caiman (*Melanosuchus niger*) *Wilderness Environ Med.* 22, 62–64.

Haddad Jr., V., Schwartz, E.F., Schwartz,C.A., Carvalho,L.N., (2010). Bites Caused by Giant Water Bugs Belonging to Belostomatidae Family (Hemiptera, Heteroptera) in Humans: A Report of Seven Cases . *Wilderness Environ. Med.* 21, 130–133.

Haddad Jr., V., Cardoso, J.L.C., Neto, D.G., (2013a). Injuries by marine and freshwater stingrays: history, clinical aspects of the envenomations and current status of a neglected problem in Brazil. *J. Venom Anim. Toxins Incl. Trop. Dis.* 19, 16.

Haddad Jr.. ,V., (2013). Environmental dermatology: skin manifestations of injuries caused by invertebrate aquatic animals. *An. Bras. Dermatol.* 88, 496-506.

Haddad,V. Jr., (2018). Injuries caused by fish in a community of Pantanal fishermen: detection, treatment, and prevention of envenomations and trauma . *Rev. Soc. Bras. Med. Trop* . 51, 700-704.

Haddara , M.M., Haberisoni ,J.B., Trelles, M., Gohou,J., Christellsa, K., Dominguez ,L., Ali,E., (2020). Hippopotamus bite morbidity: a report of 11 cases from Burundi. *Oxf .Med. Case Rep.* 2020, omaa061.

Hasan, https://scialert.net/fulltext/?doi=pjbs.2007.773.777

Hasanzadeh,M., (2019). Unusual vaginal bleeding due to a leech bite in a girl from a tropical area: A case report. *Rev. Soc. Bras. Med. Trop.* 52:e20180425

Hashim, A., Mulcahy, G., Bourke, B., Clyne, M., (2006). Interaction of *Cryptosporidium hominis* and *Cryptosporidium parvum* with Primary Human and Bovine Intestinal Cells. *Infect. Immun.* 74, 99–107.

Hashimoto, Y., Okaichi,T., Dang,L.D., Noguchi,T., (1968). Glenodinine, an Ichthyotoxic Substance Produced by a Dinoflagellate, *Peridinium polonicum. Bull. Japan. Soc. Sci. Fish.* 34, 528-534.

Hayes , A.W., (2007). *Principles and Methods of Toxicology*. CRC Press, Medical, 2296 p.

Henry, F.J., Cadiet, J., Javaudin,F., Rozec,B., *Oenantha crocata*: A Case Report of Multiple Poisoning with Fatal Outcome. *J. Emerg Med* . 2020, 59.

Hesselink, J.M.K., (2018). Kambo and its Multitude of Biological Effects: Adverse Events or Pharmacological Effects?. Int Arch Clin Pharmacol 4:017. doi.org/10.23937/2572-3987.1510017

Huang, G., Goldstein, R., Mildvan, D., (2013). Catfish spine envenomation and bacterial abscess with *Proteus* and *Morganella*: a case report. *J. Med. Case Reports* **7**, 122 .

Hudon,C., De Sève,M., (2014).Increasing occurrence of the benthic filamentous cyanobacterium *Lyngbya wollei*: a symptom of freshwater ecosystem degradation. *Antonella Cattaneo. Freshwater Science* 33, 606-618.

Ibrahim, A.M., Ahmed, A.K.,(2019). Trematode cercarial fauna obtained from the field-collected freshwater snails *Lymnaea natalensis* in Egypt. *Bull. Natl. Res. Cent*. 43, 86.

Ishii,S., Yan,T., Shively,D.A., Byappanahalli,, M.N., Whitman,, R.L., Sadowsky, M.J., (2006). *Cladophora* (Chlorophyta) spp. Harbor Human Bacterial Pathogens in Nearshore Water of Lake Michigan. *Appl. Environ. Microbiol.* 72, 4545–4553.

Jakubowska,N., Szel g-Wasielewska.E., (2015). Toxic Picoplanktonic Cyanobacteria—Review. *Mar Drugs* 13, 1497–1518.

Jaussi,R., Kunz,P.A., (1978). Isolation of the major toxic protein from the skin venom of the crested newt, *Tritutrus cristatus*. *Experientia* 34, 503–504.

Jawad,L., (2021). Ch.73. The dangerous catfish species in the freshwater system of Iraq: First time reports on cases of information. In: *Tigris and Eupharates Rivers: Their Envionments from Headwaters to Mouth*. Springer Nature, 1640p.

Junqueira, M.E.P., (2006). Bacterial contamination of the catfish (*Cathorops agassizii* - Valenciennes, 1839 and *Genidens genidens* - Agassiz, 1829) sting. *J. Venom. Anim. Toxins incl. Trop. Dis*. 12, .522.

Kaewpitoon, N., Kaewpitoon, S.J., Prasit Pengsaa,P., Sripa,B., (2008). *Opisthorchis viverrini*: The carcinogenic human liver fluke. *World J. Gastroenterol*. 14, 666–674.

Kalinda, C., Chimbari,M.J., Mukaratirwa,,S., (2017). Effect of temperature on the *Bulinus globosus — Schistosoma haematobium* system. *Infect. Dis. Poverty* 6,57.

Kim, J.H., Mok,J.S., Son,K., (2009). Toxicity of Puffer Fish, *Takifugu poecilonotus* (Heuinjeombok) and *Takifugu vermicularis* (Gukmaeribok) from Coastal Water of Korea . *Korean J. Fish. Aquat. Sci.* 42,1-7.

Kim, Y.H., Brown, G.B., Mosher, H.S.,(1975). Tetrodotoxin: occurrence in atelopid frogs of Costa Rica. *Science* 189, 151-152.

Kimura, L.F., Prezotto-Neto, J.P., Antoniazzi, M.M., Jared, S.G.S., Santoro, M.L., Barbaro, K.C., (2014). Characterization of inflammatory response induced by *Potamotrygon motoro* stingray venom in mice. *Exp. Biol. Med.*, 239, 601-609.

Kowalski , K., Marciniak , P., Rosi ski , G., Rychlik ,L., (2018). Toxic activity and protein identification from the parotoid gland secretion of the common toad *Bufo bufo*. *Comp. Biochem. Physiol. C Toxicol. Pharmacol.* 2018 205,43-52.

Kuo,H., Hsu,C., Chen,J., Wu, Y., Shen,Y., (2007). Life-threatening episode after ingestion of toad eggs: a case report with literature review. *Emerg. Med. J.* 24,215–216.

Lameiras , J. L.V., da Costa , O.T., Moroni , F.T., Araújo , J.D., Caranhas , S.M.E., Marques , C.M.A., Dos-Santos , M.C., Duncan,W.L.P., (2014). Systemic rhabdomyolysis induced by venom of freshwater stingrays *Plesiotrygon iwamae* and *Potamotrygon motoro* (Chondrichthyes-Potamotrygonidae) from the Amazon Basin. *Toxicon* 77,105-113.

Leblond, J.D., Dahmen, A.S., Lebret, K., Rengefors,K., (2013). Sterols of the Green-Pigmented, Freshwater Raphidophyte, *Gonyostomum semen*, from Scandinavian Lakes. *J. Eukaryot. Microbiol.* 60, 399-405.

Lefkowitz, R.Y., Taylor,J., Balfe,D., (2012). Reality Bites: A Case of Severe Rattlesnake Envenomation. *J. Intensive Care Med.* 28, 314-319.

Lewis,J.V., Portera Jr. C.A., (1994). Rattlesnake bite of the face: case report and review of the literature. *Am. Surg.* 60, 681-682.

Litwinowicz,A., Blaszkowska,J., (2013). *Hirudo verbana* is a source of fungal isolates potentially pathogenic to humans. *Afr. J. Microbiol. Res.* 7, 5358-5363.

Liu, H., Zhang, Z., Huang, G., Gu, X., Wang, C., Wang, Y., Lu,Z., (2017). Infection of Oligochaetes, *Limnodrilus hoffmeisteri* (Annelida: Oligochaeta), in the Nasal Cavity of a Chinese Man. *Korean J. Parasitol.* 55, 77–79.

Lok,U., Bozkurt,S., Okur,M., Gulacti,U., Hatipoglu,S., (2013). A rare case of adverse effects caused by leech bite. *Am. J. Case Rep.* 14, 191–193.

Lopez, C.B., Jewett, E.B., Dortch, Q., Walton, B.T., Hudnell, H.K., (2008). *Scientific Assessment of Freshwater Harmful Algal Blooms.* Interagency Working Group on Harmful Algal Blooms, Hypoxia, and Human Health of the Joint Subcommittee on Ocean Science and Technology. Washington, D.C., USA

Magwaza,N.M., Nxumalo,E.N., Mamba, B.B., Msagati , T.A.M., (2017). The Occurrence and Diversity of Waterborne Fungi in African Aquatic Systems: Their Impact on Water Quality and Human Health. *Int. J. Environ. Res. Public Health* 14, 546.

Mailho-Fontana, P.L., Jared, C., Antoniazzi, M.M., Sciani, J.M., Pimenta, D.C., Stokes, A.N., Grant,T., Brodie III,E.D., Brodie Jr. E.D., (2019).Variations in tetrodotoxin levels in populations of *Taricha granulosa* are expressed in the morphology of their cutaneous glands. *Sci. Rep.* 9, 18490.

Manning,S.R., Claire, J.L. (2010) Prymnesins: Toxic Metabolites of the Golden Alga, *Prymnesium parvum* Carter (Haptophyta). *Mar.Drugs* 8, 678-704.

Marciano-Cabral,F., Cabral, G., (2003). *Acanthamoeba* spp. as Agents of Disease in Humans. *Clin. Microbiol. Rev.* 16, 273–307.

Marmen ,S., Aharonovich , D., Grossowicz,M., Blank,L., Yacobi ,Y.Z., Sher,D.J.,(2016). Distribution and Habitat Specificity of Potentially-Toxic *Microcystis* across Climate, Land, and Water Use .Gradients. *Front. Microbiol.* 7, 271.

Mata,H.K., Salah,D.M.M.A., Konde,J.N., Kiyombo , G.M., Mulaji,C.K., Pote,J.W., (2020). Level of Toxic Metals in Consumable Aquatic Plant *Ledermanniella schlechteri* from Congo River and Potential Risk Assessment through Plant Consumption. *HSOA J. Food Sci.Nutr.* 6,74.

Medeiros, , J.M., Oliveira ,I.S., Ferreira ,S.G., Alexandre-Silva , G.M., Cerni,F.A., Zottich ,U., Pucca, M.B., (2020). Fatal Rattlesnake Envenomation in Northernmost Brazilian Amazon: A Case Report and Literature Overview. *Rep.- Medical Cases Images and Videos* 3, 9.

Merovich,C.E., (2005). Bufadienolides In The Chemical Defenses Of The Toads, *Bufo americanus* and *Bufo Fowleri*. https://scholarworks.wmich.edu/dissertations/1049.

Mol, https://doi.org/10.1163/9789004207653_009

Monteiro-dos-Santos J., Conceicao K., Seibert C.S., Marques E.E., Silva P.I., Jr., Soares A.B., Lima C., Lopes-Ferreira M., (2011). Studies on pharmacological properties of mucus and sting venom of *Potamotrygon* cf. *henlei. Int. Immunopharmacol.* 11,1368–1377.

Morsy , G.H., Ahmad Z Al-Herrawy ,A.Z., Elsenousy , W.M., Marouf, M.A., (2016). Prevalence of Free-Living Amoebae in Tap Water and Biofilm, Egypt. *RJPBCS* 7, 752-759.

Morton,A., Peter, J., (2015). Snakebites: 2 Cases of Suspected Crotaline Envenomation. *Consultant360,* 1, 4.

Murphey, D.K., Septimus ,E.J., Waagner. ,D.C., (1992). Catfish-Related Injury and Infection: Report of Two Cases and Review of the Literature. *Clin. Infect. Dis.* 14, 689-695.

Negreiros , M.M.B., Yamashita , S., Sardenberg , T., Fávero Junior , E.L., Ribeiro ,A.H., Haddad Junior ,W.T., Haddad Junior ,V., (2016). Diagnostic imaging of injuries caused by venomous and traumatogenic catfish . *Rev. Soc .Bras. Med. Trop.* 49, :530-533.

Nosowitz ,2017. https://www.friendsofthewildflowergarden.org/pages/plants/waterhemlock.html.

Nyakaana , S., Stothard, J.R., Nalugwa, A., Webster, B.L., Lange,C.N., Jørgensen, A., Rollinson,D., Kristensen, T.K., (2013).*Bulinus globosus* (Planorbidae; Gastropoda) populations in the Lake Victoria basin and coastal Kenya show extreme nuclear genetic differentiation. . *Acta Trop.* 128, 226-323.

Oliveira, J.S., Fernandes,S.C.R., Schwartz, C.A., Bloch Jr.C., , Melo, J.A.T., Pires Jr. O.R., Freitas, J.C., (2006).Toxicity and toxin identification in *Colomesus asellus*, an Amazonian (Brazil) freshwater puffer fish. *Toxicon* 48, 55-63.

O'Neill, M., (2021). Case Report: A Toxic Algae Bloom Might Have Killed a Family of 3 and Their Dog in California—Here's What You Need to Know. https://www.health.com/mind-body/what-is-toxic-algae-bloom.

Paerl,H.W., Fulton,R.S., Moisander,P.H., Dyble,J., (2001). Harmful Freshwater Algal Blooms, with an Emphasis on Cyanobacteria. The Scientific World 1, 76-113.

Pandey, P.K., Kass,P.H., Soupir, M.L., Biswas, S., Singh, V.P., (2014). Contamination of water resources by pathogenic bacteria. *AMB Express* 4: 51.

Patocka,J., Gupta, R.C., Kuca, K., (2011). Anatoxin-a(s): Natural organophosphorus anticholinesterase agent. *Mil. Med. Sci. Lett*.. 80, 129-139.

Patterson,G.M.L., Harris,D.O., (1983). The effect of *Pandorina morum* (Chlorophyta) toxin on the growth of selected algae, bacteria and higher plants. *Brit. Phycol. J*. 18, 259-266.

Pointier,J., Paraense,W.L., DeJong,R.J., Loker,E.S., Bargues,M.D., Mas-Coma,S., (1966).A potential snail host of schistosomiasis in Bolivia: *Biomphalaria amazonica* Paraense. *Mem. Inst. Oswaldo Cruz, Rio de Janeiro* 97, 793-796.

Qadri, S.M., al-Okaili,G.A., al-Dayel,F., (1989). Clinical significance of *Blastocystis hominis*. *J. Clin. Microbiol*. 1989 27, 2407–2409.

Ramos, H.A.C., (2017). Commercial species of freshwater stingrays in Brazil. https://cites.org/sites/default/files/eng/resources/pub/other/CITES.%20%20Freshwater_Stingrays_Sheets___2016.pdf

Rastogi R.P., Madamwar, D., Incharoensakdi, A., (2015.) Bloom dynamics of cyanobacteria and their toxins: environmental health impacts and mitigation strategies. *Front. Microbiol*. 6, 1254.

Rengefors ,K., Legrand, C,, (2007). Broad allelopathic activity in *Peridinium aciculiferum* (Dinophyceae). *Eur. J. Phycol*. 42, 341–349.

Richardson ,W.H., Barry,J.D., Tong,T.C., Williams ,S.R., Clark,R.F., (2005). Rattlesnake envenomation to the face of an infant. *Pediatr .Emerg. Care* 21,173-176.

Roberts, M.Z., McKinzie ,B.P., Leon,S.M., Fann,S.A., Bohm, N.M., (2017). Prolonged Coagulopathy Following Rattlesnake Envenomation Responsive to Antivenom: A Case Report. *Clin.Med. Case Rep*. 3.

Rodriguez-Zaragoza,S., (1994). Ecology of Free-Living Amoebae. *Crit. Rev. Microbiol*. 20, 225-241.

Roelke,D.L., Manning,S.R., (2018). Harmful Algal Species Fact Sheet: *Prymnesium parvum* (Carter) "Golden Algae": A Compendium Desk Reference. DOI: 10.1002/9781118994672.ch16q, In *Harmful Algal Blooms* (pp.629-632)

Rohrlack, T., Christoffersen,K., Kaebernick,M., Neilan,B.A., (2004). Cyanobacterial Protease Inhibitor Microviridin J Causes a Lethal Molting Disruption in *Daphnia pulicaria*. *Appl. Environ. Microbiol*. 70, 5047–5050.

Rzymski,P., Poniedz.B., (2012). Dermatotoxins synthesized by blue-green algae (Cyanobacteria). *Postepy Dermatologii I Alergologii* 29, 47-50.

Saitanu, K., Laobhripatr,S., Limpakarnjanarat, K., Sangwanloy, O., Sudhasaneya, S., Anuchatvorakul, B., Leelasitorn,S., (1991). Toxicity of the freshwater puffer fish *Tetraodon fangi* and *T. palembangensis* from Thailand.. *Toxicon* 29, 895-897.

Saitanu,K., (1992). Toxicity of Thai freshwater pufferfish *Tetraodon nigroviridis* and *Teraodon steindachneri*. *Asian Fish. Sci* . 5, 383-386.

Sarmiento , B.E., Rangel , M., a Gonçalves , J.C., Pereira , L., Rego , S., Campos , L.A., Haddad Jr , V., Mortari ,M.R., Schwartz,E.F., (2015).First report of the characterization of the pathophysiological mechanisms caused by the freshwater catfish *Pimelodus maculatus* (order: Siluriformes). *Toxicon* 101,55-62.

Sato, S.,Kodama, M., Ogata, T., Saitanu, K., Furuya, M., Hirayama, K., Kakinuma, K,., (1997). Saxitoxin as a toxic principle of a freshwater puffer, *Tetraodon fangi*, in Thailand. *Toxicon* 35, 137-140.

Sazima , I., Zuanon ,J., Haddad,V. Jr, , (2005). Puncture Wounds by Driftwood Catfish During Bucket Baths: Local Habits of Riverside People and Fish Natural History in the Amazon. *Wilderness Environ. Med* . 16, 204 208.

Scheid, P.L., (2019). *Vermamoeba vermiformis* - A Free-Living Amoeba with Public Health and Environmental Health Significance. *Open Parasitol. J.* 7, 40-47.

Seifert, S.A., Ruha,M., (2022). Bites by Crotalinae snakes (rattlesnakes, water moccasins [cottonmouths], or copperheads) in the United States: Clinical manifestations, evaluation, and diagnosis. https://www.uptodate.com/contents/bites-by-crotalinae-snakes-rattlesnakes-water-moccasins-cottonmouths-or-copperheads-in-the-united-states-clinical-manifestations-evaluation-and-diagnosis ,

Sesen,R., (2004). A Study on the Conchological Features of Two Populations of *Bulinus truncatus* (Gastropoda: Pulmonata) in Turkey. *Turk. J. Zool*, 28 , 145-148.

Silva , G.C., Sabino, J., Alho, C.J.R., Nunes, V.L.B., Haddad Jr. V., (2010). Injuries and envenoming by aquatic animals in fishermen of Coxim and Corumbá municipalities, state of Mato Grosso do Sul, Brazil: identification of the causative agents, clinical aspects and first aid measures. *Rev .Soc .Bras. Med. Trop.* 43, 486-490.

Sivonen, K., Jones,G., (1999). Chapter 3. Cyanobacterial Toxins. Toxic Cyanobacteria in Water: A guide to their public health consequences, monitoring and management . https://www.who.int/water_sanitation_health/resourcesquality/toxcyanchap3.pdf

Smith,F.M.J., Wood,S.*A.*, *Wilks ,T.*, Kelly, D., Broady, P.A., Williamson, W., Gaw,S., (2012). Survey of Scytonema (Cyanobacteria) and associated saxitoxins in the littoral zone of recreational lakes in Canterbury, New Zealand. *Phycologia* 51, 542–551.

Sodre,J.G., Colaço- fernandes ,T.R., Py-Daniel, L.H.R., Birindelli ,J.L.O., Zuznon,J., (2021). Putative poison gland in the thorny catfish *Acanthodoras spinosissimus* (Siluriformes: Doradidae) *Acta Amazonica* 51, 250 – 254.

Sotton,B., Domaizon,I., Anneville ,O., Cattane´o,F., Guillard,J., (2014). Nodularin and cylindrospermopsin: a review of their effects on fish. *Rev. Fish. Biol. Fisheries* 25, DOI: 10.1007/s11160-014-9366-6

Tanu, M.B., Mahmud, Y., Tsuruda, K., Arakawa, O., Noguchi, T., (2001). Occurrence of tetrodotoxin in the skin of a rhacophoridid frog *Polypedates* sp. from Bangladesh. *Toxicon* 39, 937-941.

Thackston ,D., Wills,B.K., (2022). Water Moccasin Snake Toxicity. Treasure Island (FL): StatPearls Publishing. https://www.ncbi.nlm.nih.gov/books/NBK546645/

Thalgaspitiya,S.P.B., Wijerathne ,B.T.B., Siriwardhana, , N.P., (2017). Freshwater Catfish Envenoming in a Tropical Country. *Wilderness Environ. Med.* 28, 259–266.

Thomas , V., Loret , J., Jousset, M., Greub, G., (2008). Biodiversity of amoebae and amoebae-resisting bacteria in a drinking water treatment plant. *Environ. Microbiol.* 2008, 10, 2728-2745.

Toledo, R.C., Jared,C., (1995). Cutaneous granular glands and amphibian venoms. *Comp. Biochem. Physiol.* 111, l-29.

Torrez, P.P.Q., Quiroga, M.M., Said,R., Abati, P.A.M., Franca, F.O.S., (2015). Tetanus after envenomations caused by freshwater stingrays. *Toxicon* 97, 32-35.

Trevino-Garrison ,K.I., DeMent ,J., Ahmed, F.S., Haines-Lieber ,P., Langer ,T., Ménager , H., Neff , J., van der Merwe, D., Carney,E., (2015). Human Illnesses and Animal Deaths Associated with Freshwater Harmful Algal Blooms. *Toxins* 7, 353-366.

Turner, A.D., Dhanji-Rapkova,M., Neill,A.O., Coates,L., Lewis,A., Lewis,K., (2018). Analysis of Microcystins in Cyanobacterial Blooms from Freshwater Bodies in England. *Toxins* 2018, 10, 1-29.

USDA Agricultural Research Service. https://www.ars.usda.gov/pacific-west-area/logan-ut/poisonous-plant-research/docs/water-hemlock-cicuta-douglasii/

Usman, B.I., Amin, S.M.N., Arshad A, Rahman, M.A.,(2013). Review of some biological aspects and fisheries of Grey-Eel Catfish *Plotosus canius* (Hamilton, 1822). *Asian J. Anim. Vet .Adv.* 2,154–167.

Venkaiah ,Y., Lakshmipathi, V., (2000). Biochemical Composition of Epidermal Secretions and Poisonous Spine of Two Freshwater Catfishes. *Asian Fish. Sci.* 13, 183-189.

Violi ,J.P., Facey, J.A., Mitrovic,S.M., Colville,A., Rodgers.,K.J., (2019). Production of β-methylamino-L-alanine (BMAA) and Its Isomers by Freshwater Diatoms. *Toxins* 11, 512.

Volkmer-Ribeiro ,C., Lenzi , H.L., OrÈfice ,F., Pelajo-Machado , M., de Alencar , L.M., Fonseca ,C,F., Batista ,T.C.A., Manso ,P.D.A., Coelho ,J., Machado ,M., (2006). Freshwater sponge spicules: a new agent of ocular pathology. *Mem .Inst. Oswaldo Cruz, Rio de Janeiro*, 101, 899-903.

Vyas , K.J., Weiss, J.H., (2009). BMAA--an unusual cyanobacterial neurotoxin. *Amyotroph Lateral Scler.* 2009 10, 50-55.

Walker, A.A., Weirauch, C., Fry,B.G., King,G.F., (2016). Venoms of Heteropteran Insects: A Treasure Trove of Diverse Pharmacological Toolkits. *Toxins* 2016, 8, 43.

Wannasan , A., Uparanukraw , P., Songsangchun,A., Morakote, N., (2013).Potentially pathogenic free-living amoebae in some flood-affected areas during *2011 Chiang Mai flood.* . *Rev. Inst. Med. trop. S. Paulo* 55, 411-416.

Watt, G., Theakston,R.D., (1985). Seasnake bites in a freshwater lake. Am. J. Trop. Med. Hyg. 34,770-773.

Welzel, G., Schuster, S., (2021). Efficient high-voltage protection in the electric catfish . *J. Exp. Biol.* 224, jeb239855.

WHO, (1998). Guidelines for drinking water quality. Vol.2. *Health criteria and other supporting information*

WHO, (2019). Cyanobacterial toxins: Anatoxin-a. https://www.who.int/water_sanitation_health/water-quality/guidelines/chemicals/anatoxin-a-gdwq-bd-for-review-20191122.pdf

Wiese, M., D'Agostino,P.M., Mihali,T.K., Moffitt,M.C., Neilan, B.A., (2010). Neurotoxic Alkaloids: Saxitoxin and Its Analogs. *Mar. Drugs* 2010; 8, 2185–2211.

Williams, B.L., Powers, L.V., Garner,M.M., (2016). A pufferfish (*Tetradon nigroviridis*) available in the common pet trade harbors lethal concentrations of tetrodotoxin: A case study of poisoning in a Cuvier's dwarf caiman (*Paleosuchus palpebrosus*). *J. Zoo. Wildl. Med.* 2016 47, 676-680.

Wright,J.J., (2009). Diversity, phylogenetic distribution, and origins of venomous catfishes. *BMC Evol. Biol.* 2009, 9,282.

Wright,J.J., (2012). *The evolutionary ecology of venomous catfishes, with a focus on members of the North American family Ictaluridae (Teleostei: Siluriformes).* Ph.D.thesis, University of Michigan.

Wright,J.J., (2012). Evolutionary History of Venom Glands in the Siluriformes. In: *Evolution of Venomous Animals and Their Toxin.* Springer Science, 19p.

Xiao , L., Fayer, R., Ryan, U., Upton,S.J., (2004). *Cryptosporidium* Taxonomy: Recent Advances and Implications for Public Health. *Clin. Microbiol. Rev.* 2004, 72–97.

Yasumoto, T., Yotsu, M., Murata, M. , (1988). New tetrodotoxin analogues from the newt *Cynops ensicauda. J. Am. Chem. Soc.* 110, 2344-2345.

Yatigammana, S.A., Ileperuma, O.A., Perera,M.B.U., (2011). Water pollution due to a harmful algal bloom: a preliminary study from two drinking water reservoirs in Kandy, Sri Lanka. *J.Natn..Sci..Foundation Sri Lanka* 39 , 91-94.

Yohannan,B., Feldman, M., (2019). Fatal *Balamuthia mandrillaris* Encephalitis. *Clin. Case Rep. Infect. Dis.* 2019, 9315756.

Yotsu- -Yamashita, M., Mebs, D. , (2001). The levels of tetrodotoxin and its analogue 6-epitetrodotoxin in the red-spotted newt *Notophthalmus viridescens. Toxicon* 39, 1261-126.

Zimba, P.V., Huang ,I., Gutierrez ,D., Shin ,W., Bennett , M.S., Triemer, R.E., (2017). Euglenophycin is produced in at least six species of euglenoid algae and six of seven strains of *Euglena sanguinea*. *Harmful Algae* 63, 79-84.

Web References

file:///C:/Users/rsant/Downloads/160727_Reveillon_revisedarticle_archimer.pdf

http://txstate.fishesoftexas.org/noturus%20gyrinus.htm

http://www.cdc.gov; https://mcdinternational.org/trainings/malaria/english/dpdx5/HTML/ImageLibrary/A-F/FreeLivingAmebic/body_FreeLivingAmebic_il5

http://www.duketownstingrays.com/Files/2/60000/60267/Attachments/Product/49zdG998465t182d7098g040k042cv16.pdf

http://www.t3db.ca/toxins/T3D2528

http://www.wikitox.org/doku.php?id=wikitox:penetrating_marine_envenoming

https://adventure.howstuffworks.com/top-5-poisonous-plants10.htm

https://amphibiaweb.org/cgi/amphib_query?where-genus=Oophaga&where-species=granulifera&account=amphibiaweb

https://amphibiaweb.org/cgi/amphib_query?where-genus=Phyllobates&where-species=bicolor

https://amphibiaweb.org/species/110

https://amphibiaweb.org/species/1704

https://amphibiaweb.org/species/1706

https://amphibiaweb.org/species/1708

https://amphibiaweb.org/species/4559

https://animaldiversity.org/accounts/Agkistrodon_piscivorus/

https://animaldiversity.org/accounts/Alligator_mississippiensis/

https://animaldiversity.org/accounts/Atractosteus_spatula/

https://animaldiversity.org/accounts/Carcharhinus_leucas/

https://animaldiversity.org/accounts/Crocodylus_acutus/

https://animaldiversity.org/accounts/Crotalus_horridus/; https://reptile-database.reptarium.cz/species?genus=Crotalus&species=horridus

https://animaldiversity.org/accounts/Dendrobates_auratus/

https://animaldiversity.org/accounts/Dendrobates_leucomelas/

https://animaldiversity.org/accounts/Himantura_chaophraya/

https://animaldiversity.org/accounts/Hippopotamus_amphibius/

https://animaldiversity.org/accounts/Hirudo_medicinalis/

https://animaldiversity.org/accounts/Malapterurus_electricus

https://animaldiversity.org/accounts/Melanosuchus_niger/https://animaldiversity.org/accounts/Alligator_sinensis/

https://animaldiversity.org/accounts/Noturus_exilis/

https://animaldiversity.org/accounts/Sistrurus_miliarius/

https://animaldiversity.org/accounts/Vandellia_cirrhosa/

https://askinglot.com/are-freshwater-leeches-harmful

https://australianmuseum.net.au/learn/animals/worms/leeches/

https://a-z-animals.com/blog/how-many-people-do-cottonmouths-water-moccasins-bite-per-year/

https://carteret.ces.ncsu.edu/wp-content/uploads/2013/05/Poisonour-Plants-of-the-Southern-United-States.pdf?fwd=no

https://crsreports.congress.gov/product/pdf/R/R44871,2020

https://davesgarden.com/guides/pf/go/1172/#b

https://en.wikipedia.org/wiki/Acer_rubrum

https://en.wikipedia.org/wiki/American_crocodile

https://en.wikipedia.org/wiki/Atelopus

https://en.wikipedia.org/wiki/Biomphalaria_sudanica

https://en.wikipedia.org/wiki/Corroboree_frog

https://en.wikipedia.org/wiki/Eastern_copperhead

https://en.wikipedia.org/wiki/Fahaka_pufferfish

https://en.wikipedia.org/wiki/Fire_salamander

https://en.wikipedia.org/wiki/Lily_of_the_valley

https://en.wikipedia.org/wiki/Mugger_crocodile

https://en.wikipedia.org/wiki/Naja_annulata

https://en.wikipedia.org/wiki/Naja_christyi

https://en.wikipedia.org/wiki/Nile_crocodile

https://en.wikipedia.org/wiki/Nodularia

https://en.wikipedia.org/wiki/Oenanthe_crocata

https://en.wikipedia.org/wiki/Phyllomedusa_bicolor

https://en.wikipedia.org/wiki/Pseudophryne

https://en.wikipedia.org/wiki/Red-backed_poison_frog

https://en.wikipedia.org/wiki/Rough-skinned_newt

https://en.wikipedia.org/wiki/Sappinia

https://en.wikipedia.org/wiki/Smooth_toadfish

https://en.wikipedia.org/wiki/Timber_rattlesnake#Distribution_and_habitat

https://en.wikipedia.org/wiki/Toxoplasma_gondii

https://en.wikipedia.org/wiki/Zimmerman%27s_poison_frog

https://fishesofaustralia.net.au/home/species/2762

https://fishesofaustralia.net.au/home/species/878

https://hollowaysbeacheec.eq.edu.au/SupportAndResources/FormsAndDocuments/Documnts/fish-toadfish.pdf

https://in.mashable.com/science/6676/researchers-have-found-the-worlds-most-powerful-electric-eel-in-the-amazon

https://insmoothwaters.com/most-dangerous-freshwater-fish/#:~:text=The%20goonch%20fish%20is%20also,as%20a%20predator%20to%20humans

https://mywaterquality.ca.gov/habs/images/algae_mitigation_technique_seletion_process_for_lakes.jpg

https://news.mongabay.com/2005/08/study-discovers-why-poison-dart-frogs-are-toxic/

https://okinawanaturephotography.com

https://piranhaguide.com/which-piranha-is-the-most-aggressive/

https://plants.ces.ncsu.edu/plants/nerium-oleander

https://plants.ces.ncsu.edu/plants/pistia-stratiotes/

https://plants.ces.ncsu.edu/plants/rhododendron-maximum/

https://plants.ces.ncsu.edu/plants/solanum-dulcamara/#poison

https://pondinformer.com/are-catfish-dangerous/

https://reptile-database.reptarium.cz/species?genus=Laticauda&species=crockeri) with the presence of a short neurotoxin B (https://www.uniprot.org/uniprot/P25496

https://reptile-database.reptarium.cz/species?genus=Naja&species=annulata

https://revistapesquisa.fapesp.br/en/2013/04/02/the-dangers-of-swimming-in-the-araguaia

https://rivers.dwer.wa.gov.au/species/tandanus-bostocki/#:~:text=The%20mouth%20is%20surrounded%20by,venom%20gland%20cells%20is%20torn

https://roaring.earth/these-are-the-deadliest-freshwater-fish/

https://shodhganga.inflibnet.ac.in/bitstream/10603/151800/14/14_chapter%206.pdf

https://whatdewhat.com/dangerous-freshwater-fish-can-kill/

https://wiki.bugwood.org/Iris_pseudacorus

https://wwf.panda.org/discover/our_focus/wildlife_practice/profiles/fish_marine/stingray/

https://www.acep.org/tox/case-files/27_Rattlesnake%20Envenomation%202.pdf

https://www.aquariumglaser.de/en/fish-archives/malapterurus-beninensis-electric-catfish/

https://www.aquariumglaser.de/en/fish-archives/thalassophryne-amazonica-4/

https://www.aquasabi.com/Hygrophila-balsamica-pot

https://www.britannica.com/animal/leech#ref57162

https://www.britannica.com/science/Cladophora#ref1278905

https://www.britannica.com/science/steroid/Toad-poisons#ref100185

https://www.cabi.org/isc/datasheet/121720

https://www.cdc.gov/dpdx/opisthorchiasis/index.html

https://www.cdc.gov/parasites/clonorchis/index.html

https://www.cdc.gov/parasites/naegleria/index.html

https://www.cdc.gov/parasites/sappinia/general.html

https://www.ch.ic.ac.uk/local/projects/assavathorn/Tetrodotoxin.html

https://www.des.nh.gov/organization/commissioner/pip/factsheets/bb/documents/bb-10.pdf

https://www.dixieanimalhospital.com/blog/13014-beware-toads-that-can-poison-cats-and-dogs

https://www.ebmedicine.net/topics/toxicology-environmental/snake-bite-antivenom

https://www.ehow.com/list_6628387_poisonous-water-plants.htm

https://www.fishbase.de/summary/acanthodoras-spinosissimus.html

https://www.fishbase.de/summary/plotosus-papuensis.html

https://www.fishbase.de/summary/Pygocentrus-nattereri.html

https://www.fishbase.de/summary/Takifugu-rubripes.html

https://www.fishbase.in/summary/12095

https://www.fishbase.in/summary/51770

https://www.fishbase.in/summary/68307

https://www.fishbase.in/summary/Porochilus-argenteus.html

https://www.fishbase.in/summary/Takifugu-obscurus.html._

https://www.fishbase.se/summary/13394

https://www.fishbase.se/summary/3054

https://www.fishbase.se/summary/53559

https://www.fishbase.se/summary/Cathorops-agassizii.html

https://www.fishbase.se/summary/Cathorops-aguadulce

https://www.fishbase.se/summary/Clarias-fuscus

https://www.fishbase.se/summary/Neosilurus-ater.

https://www.fishbase.se/summary/Tandanus-tandanus

https://www.futurity.org/just-your-garden-variety-poisonous-catfish

https://www.fws.gov/fisheries/ans/erss/uncertainrisk/ERSS_Heteropneustes_microps_final_January_2018.pdf

https://www.gardenguides.com/127812-poisonous-water-plants.html

https://www.health.com/condition/infectious-diseases/flesh-eating-bacteria-long-island-sound

https://www.healthline.com/health/rattlesnake-bite#symptoms

https://www.kuow.org/stories/a-new-bloodsucking-leech-species-found-hiding-outside-washington-d-c

https://www.medicalnewstoday.com/articles/323689

https://www.myaquariumclub.com/are-clarias-catfish-poisonousi-have-read-online-that-there-are-some-cases-...-507056.html

https://www.nationalgeographic.com/animals/fish/e/electric-eel/

https://www.nationalgeographic.com/news/2015/03/150311-giant-stingray-largest-freshwater-fish-thailand-whiptail-animals/

https://www.nationalgeographic.org/media/photo-ark-green-spotted-pufferfish

https://www.nchmd.org/education/mayo-health-library/details/CON-20155618

https://www.petpoisonhelpline.com/blog/copperhead-snakes-envenomation-risk-for-animals/

https://www.poison.org/articles/poisoning-from-amphibians-204#:~:text=Some%20of%20the%20potent%20toxins,to%20spread%20Salmonella%20to%20humans

https://www.practicalfishkeeping.co.uk/features/definitive-guide-to-south-american-freshwater-rays/

https://www.practicalfishkeeping.co.uk/features/piraya-piranha-pygocentrus-piraya/

https://www.researchgate.net/profile/Alaa_Eissa2/publication/303005861

https://www.scielo.br/scielo.php?script=sci_arttext&pid=S1678-91992006000200001#co

https://www.scienceabc.com/nature/animals/diving-bell-spider-anatomy-habitat-bite-facts.html

https://www.sciencealert.com/here-s-a-list-of-the-top-15-deadliest-animals-on-earth

https://www.sciencedirect.com/topics/nursing-and-health-professions/nodularin

https://www.scotcat.com/articles/article119.html

https://www.scotcat.com/factsheets/malapterurus_microstoma.htm

https://www.tampabay.com/news/latest/2022/06/22/boy-stabbed-by-catfish-in-new-port-richey-expected-to-fully-recover/

https://www.thelocal.de/20160825/6-foot-catfish-attacks-woman-in-bavarian-lake/

https://www.timesnownews.com/the-buzz/article/newly-discovered-electric-eel-is-the-most-powerful-can-generate-860-volt-shock/488043

https://www.treehugger.com/beautiful-deadly-frogs-4859270

https://www.wbtv.com/2019/10/16/venomous-fish-with-painful-sting-was-presumed-extinct-smoky-mountains-its-back

INDEX

D